Physics of Biological Action and Perception

Physics of Biological Action and Perception

Mark L. Latash

Department of Kinesiology,
The Pennsylvania State University,
University Park, PA, United States

ELSEVIER

ACADEMIC PRESS

An imprint of Elsevier

Academic Press is an imprint of Elsevier
125 London Wall, London EC2Y 5AS, United Kingdom
525 B Street, Suite 1650, San Diego, CA 92101, United States
50 Hampshire Street, 5th Floor, Cambridge, MA 02139, United States
The Boulevard, Langford Lane, Kidlington, Oxford OX5 1GB, United Kingdom

Library of Congress Cataloging-in-Publication Data
A catalog record for this book is available from the Library of Congress

British Library Cataloguing-in-Publication Data
A catalogue record for this book is available from the British Library

ISBN: 978-0-12-819284-9

Publisher: Nikki Levy
Acquisition Editor: Melanie Tucker
Editorial Project Manager: Tracy I Tufaga
Production Project Manager: Selvaraj Raviraj
Cover Designer: Miles Hitchen

Typeset by TNQ Technologies

Contents

Preface

In this book, I have tried to summarize the lifetime experience in using the logics of natural science (physics) to address problems inherent to biological systems. Why physics? Because "Only physics is the salt, all the rest are nil" as we used to sing in the 1970s as undergraduate students in the Moscow Institute of Physics and Technology, my *alma mater*. And the song continued: "And the philosopher and biologist are bludgeons." The last word in Russian has two meanings, the direct one and "a hopelessly dumb person" I hope that biologists and philosophers forgive me for this quotation (I have not made it up!), particularly because among my physicist friends, I have the reputation of a biologist and a bit of a philosopher.

A few great scientists of the past made statements highly relevant to the spirit of the book, including two Nobel Prize winners in physics. A number of sources claim that Ernest Rutherford (1871–1937), who is sometimes addressed as the father of nuclear physics, said "*All science is either physics or stamp collecting*." (There are many references to this quotation, albeit all of them are somewhat imprecise. Probably, the most reliable one is in the classical book by J.D. Bernal, "*The Social Function of Science*" published in 1939 and republished by Faber and Faber in 2010.) Indeed, collecting stamps may be fun, nearly as much fun as performing scientific research or even more, and one can apply tools provided by mathematics to analyze stamps. For example, after painstakingly meticulous analysis, one might be able to prove that, statistically, red stamps are more expensive than green ones and that triangular stamps are rare and usually have larger area than square ones.

The other quotation is from Erwin Schrödinger (1887–1961), one of the founders of quantum physics and the author of a classic "*What is Life? The Physical Aspect of the Living Cell*" published in 1944: "*If you suspend laws of nature in the human body, you can explain anything*." This quotation is from the 1949 lecture "Do Electrons Think?" (what a great title!), which can be found on the Internet: https://www.youtube.com/watch?v=hCwR1ztUXtU. I am very much indebted to my friend as well as colleague, Sandro Mussa-Ivaldi, for pointing me toward this great lecture. I hope that all readers appreciate sarcasm of the quoted phrase: the word "anything" can be replaced with "nothing" without a change in the meaning.

The third quotation comes from one of my informal mentors, a great mathematician, Israel Gelfand (1913–2009). I describe the history of this quotation in more detail in the Preamble, but now let me rephrase it without much explanation: "*We should not be afraid of saying with respect to some problems: 'These are beyond our comprehension'*." I view this statement as very important methodological guidance in my own research and recommend it to all of my students and younger colleagues.

Two of the main goals of this book have been to avoid both traps: (1) turning study of biological problems into "stamp collecting" and (2) "explaining anything" by ignoring laws of nature. The third goal was not to be afraid to admit that some problems remain hopelessly "beyond my comprehension." I am even using an abbreviation "BMC" (beyond my comprehension) for this phrase in quite a few parts of the book.

Unfortunately, we know very little about biological objects, in particular, such complex ones as the human body, and even less about relevant laws of nature. Of course, laws of classical physics apply to biological objects, but the laws we know seem only to impose constraints on possible actions by animals, but they do not define what animals actually do. In other words, our current understanding of the underlying physics is rather poor. This means that we have to search for biology-specific laws of nature, discover them, and try to understand how they emerge and what role they play in different aspects of biological behavior. To me, this obvious gap in our current knowledge is a major source of excitement. Indeed, it is great to work at the dawn of a particular area of science and, as far as physics of biological processes is concerned, we are still at a pre-Galilean stage.

The book starts with analysis of motor actions, which has been my main field of work over the past >40 years. Motor actions are very attractive to an experimentalist who tries to explore laws of nature because movements are relatively easy to quantify objectively with the help of the apparatus of classical mechanics. Of course, the "ease" is only relative because many relevant parameters and variables are not directly observable. For example, there are no tools to measure muscle forces and joint torques directly; these variables have to be estimated based on equations of motion that include many parameters known only approximately. In fact, the situation is much more complex because biological systems are active. This means, in particular, that they commonly react to measuring manipulations by changing those very parameters that the researcher tries to estimate. Hence, experimentalists have to be inventive and even sneaky to perform measurements that are all but undetected by the object of study.

The next part of the book deals with perception, which is not my direct area of expertise. So, I ask for forgiveness from my colleagues who work directly in the field of perception and may be appalled by my errors. On the other hand, I feel that one has to address this major function of the human body based on laws of nature. Thinking in that direction led to the formulation of an alternative definition of "percepts" and allowed making predictions for experimental studies that have not been performed yet. To me, this is a very promising sign suggesting that the approach is not simply a reformulation of known facts with a new vocabulary but a potentially productive direction of research.

Closer to the end of the book, I dare to venture into fields that are nearly 100% BMC, such as language communication, playing chess, and cognition in general. I have been fascinated by these areas for years and even made a few attempts at performing experimental studies. These fields remain, however, much closer to philosophy than to experiment-based science.

Finally, I end the book with a set of problems and conjectures, which I see as central to the nascent field of physics of living systems. The conjectures range from several, where I feel rather confident, to a few wild guesses.

The book assumes little preexistent knowledge in the mind of the reader, rather open-mindedness and inquisitiveness. On the other hand, knowing a little bit of physics and math, and also something about the human body (basic anatomy and physiology), would not hurt. I see a perfect reader as a freshman graduate student with background in one or more of the following areas: physics, math, physiology, physical therapy, and similar fields. But, of course, I would be happy to have anybody read this book, from inquisitive high school students to my respected colleagues. The book was not designed as a textbook, but some of my colleagues may find it useful for teaching courses at high undergraduate or graduate levels.

And now, it is time for the part that I enjoy writing more than anything else. This book would not have been written if not for the influence, help, and good attitude from many people. I would like to start with mentioning my late parents: my father, Lev Latash, was a physiologist and is among the very few scientists who coauthored papers with Nikolai Bernstein, and my mother, Sara Fuchs-Latash, was a radiologist and, more importantly, the kindest, ever-supporting person in my life. My father was my lifetime mentor, but at different stages of career I had other mentors, formal and informal, who influenced my thinking greatly. I want to mention with much warmth and gratitude Anatol Feldman, Israel Gelfand, Gerry Gottlieb, Victor Gurfinkel, and Vladimir Zatsiorsky. Anatol and Vladimir, as well as their wives, Mindy Levin and Rita Zatsiorsky, have also been very close personal friends.

I am also deeply grateful to all the graduate students, postdoctoral fellows, and visiting researchers who worked in the Motor Control Laboratory at Penn State (MCL@PSU) over the past quarter of a century where many studies cited in this book were performed. I would like to mention personally those brave 20 who have worked successfully toward doctoral degrees in the MCL@PSU over the past years: Cristian Cuadra, Alessander Danna-dos-Santos, Ali Falaki, Stacey Gorniak, Hang Jin Jo, Vijaya Krishnamoorthy, Sheng Li, Shirin Madarshahian, Daniela Mattos, Halla Olafsdottir, Sasha Reschechtko, Jae-kun Shim, Takako Shiratori, Tarkeshwar Singh, Varadhan Srinivasan Kariyamaanikam (SKM), Harmen Slijper, Yen-Hsun Wu, Yang Xu, Wei Zhang, and Tao Zhou.

I cannot possibly mention all my friends as well as colleagues who came to Penn State to present at our seminar series "Action Club" which is now in its 22nd year, and those who presented at the Annual Motor Control Summer School, which has been running over the past 17 years. I am very grateful for having learned a lot from you all, even if your names are not mentioned in this brief Preface. The names I want to mention are those of Slobodan Jaric and John Scholz, two very close personal friends and great scientists who passed away not that long ago leaving behind a void in the hearts of all people who had been lucky to know them. And, last but not least, my deep gratitude and love goes to my wife Irina, my daughters Liza Jr. and Liza Sr., and my son Samvel.

The preamble

I am forever indebted to Israel Gelfand, one of the greatest mathematicians of the 20th century. During the later period of his life, Izrail' Moiseevich (as he was properly addressed by students and colleagues who knew Russian) allowed me to visit him in his house in New Brunswick, New Jersey, and discuss problems related to my research as well as many other, seemingly unrelated, issues (for example, whether Shostakovich performed his piano concertos better than Richter). This was an amazing mind-cleansing experience! Fortunately, I was mature enough to appreciate this unique opportunity. During one of the visits, when we seemed to get stuck in the discussion on problems of the neural control of movement, Gelfand suddenly said: "We should be able to solve this problem because we are Jews. This means that we are not afraid of saying with respect to some questions: *These are beyond our comprehension*." And he continued: "Let us write a book with the title *Beyond Our Comprehension*." Unfortunately, we did not write such a book. Only now, about 20 years later, I feel ready to attempt putting together a line of thought that was, directly or indirectly, stimulated by those conversations with Israel Gelfand. I dedicate this book to his memory.

Introduction

1

Many years ago, a great scientist, Nikolai Bernstein, decided to write a book about dexterity (Bernstein, 1996). Although all people intuitively understand what *dexterity* is, to use this term in a productive way that allows performing experimental and theoretical scientific inquiry, one has to define it. Bernstein spent the whole first chapter of the book building a definition of dexterity. He had two basic requirements for such a definition. First, it had to be operational. This means that the definition could be used to identify, or—even better—quantify, dexterity and distinguish it from "non-dexterity." Second, the definition had to match the preexistent intuitive understanding of this word as closely as possible.

Can one build a definition for *living system* using a similar strategy? Having such a definition seems crucial for providing a foundation for many established sciences such as biology, psychology, neuroscience, biomechanics, etc. Of course, one can find many definitions of *living systems* (or simply *life*) in books, dictionaries, and on the Internet. There is a problem, however. Most of these definitions do not define living systems but rather enumerate their features such as adaptation, growth, homeostasis, metabolism, reproduction, separation from the environment (e.g., by the skin or by the cellular membrane), and some others. I purposefully put this list of features in the alphabetical order not to create an impression that there are more important and less important features. Other definitions use vague terms such as *self-organization*, which themselves need a formal definition. For example, does a growing crystal self-organize? It seems to do this. But most people would not claim that a growing crystal is a living system. It self-organizes in too a predictable way to qualify as living. Still other definitions focus on a particular feature of living systems, for example, their seeming violation of the Second Law of Thermodynamics and reduction of entropy. Of course, this reduction is local and achieved at the expense of the external world.

Let us start building a definition for a living system with introspection: When we see a moving object, typically it is relatively easy to distinguish a living object from a non-living one. Inanimate objects move in predictable ways: If one knows or can guess salient parameters of an inanimate object, its initial state, and external forces, it is possible to predict its motion with high certainty. In contrast, living objects move in a much less predictable way. In particular, they commonly walk or crawl uphill, fly against the wind, and swim against the current. In other words, they are *active*. This word does not mean, of course, that living objects violate basic laws of nature. At least, we have no compelling evidence for such violations. Nikolai Bernstein appreciated the fact that *activity* was the primary distinguishing feature of biological objects. He spent the last years of his life trying to create a new science, *physiology of activity* (Bernstein, 1966). Important insights into the foundations of activity, as a defining feature of living systems, were also made by a great

Physics of Biological Action and Perception. https://doi.org/10.1016/B978-0-12-819284-9.00001-2

mathematician Israel Gelfand and a brilliant physicist Michael Tsetlin (Gelfand and Tsetlin, 1962, 1966).

Let me use a more intuitive example. Consider a living animal, e.g., a frog, and an exact replica of that animal—a toy frog. These two objects behave very differently even if all their parameters are matched perfectly given our current level of knowledge and technical sophistication. The toy is very predictable in its motion, whereas the frog obeys only the famous Harvard's Law: "Under the most rigorously controlled conditions of pressure, temperature, volume, humidity, and other variables, the organism will do as it well damn pleases" (Bloch, 2003).

There is another distinguishing, and commonly overlooked, feature of living systems: They show *stability* of important characteristics of behavior while other characteristics may be highly variable, even noisy. Here, the word *behavior* is used in a very general sense, from motor action to perception, information exchange, and cognitive processes. Consider the following few examples.

When a person performs an action multiple times, typically, the important action characteristics show relatively reproducible trajectories, while characteristics of elements contributing to the action are much more variable. This was demonstrated by Bernstein about 100 years ago in his, now famous, study of professional blacksmiths (Bernstein, 1930). He asked his subjects to perform one of their typical labor movements—hitting the chisel held in the left hand with the hammer moved by the right hand (they were right-handed)—multiple times and recorded the kinematics of the joints moving the hammer as well as of the hammer itself. Note that these were perfectly trained subjects for this action, given that they had performed it hundreds of times a day for years. Bernstein reported that the intertrial variability of the hammer trajectory was much smaller than the intertrial variability of individual joint trajectories. Clearly, this is not a trivial observation. The brain cannot send signals directly to the hammer; it can only send signals to muscles that cause joint rotations. So, how can it be that the signals to muscles produced highly variable joint trajectories while the hammer trajectory was much more consistent across trials? It took researchers over half a century to come up with a feasible explanation for this phenomenon.

Imagine now that you walk along an art gallery and move your head to look at the paintings. During this action, all sensory signals informing on the position of the head and body in space—visual, vestibular, and somatosensory—change. Nevertheless, you have an adequate, stable perception that the external world is stationary. This perception stays stable even if your motion along the gallery is not self-produced, e.g., if you sit in a wheelchair and another person pushes you along the gallery.

Consider another very simple example: Imagine that you press with the palm against a stop without moving the arm. When you vary the force, signals from all relevant peripheral sensory endings (receptors) change due to changes in the muscle geometry, tendon force, skin deformation, joint capsule tension, activity of certain spinal neurons (e.g., gamma motor neurons), etc. However, all these changes fail to violate the stable, veridical perception that the arm does not move.

Imagine now that you describe to a group of friends an important episode of your life multiple times. One does not have to run a formal experiment to agree that multiple accounts of the same episode will vary in the composition, structure of individual phrases, vocabulary, prosody, etc. However, assuming that you try to convey the important features of the episode in each of the accounts, gist of the story will remain stable despite the variable means of its delivery.

Finally, if you are asked to perform a typical everyday task multiple times, for example, to go to a grocery store a few blocks away and buy several items on the shopping list, more likely than not, you will use different trajectories from the home to the store and inside the store to accomplish the task. Some of the variability of the trajectories may be due to external factors such as streetlights, cars, pedestrians, other customers in the store, lines to the cashiers, etc. But even if the streets were empty and you were the only customer, likely your trajectories would differ spontaneously, due to unpredictable changes in your intrinsic states such as thoughts. Despite all this variability, most likely you will end up with the same (or not very variable) set of products in the shopping bag in front of your home entrance door.

What is so special about stability of the outcome in all these examples? And why doesn't the system ensure the same stable outcome by using the same, similarly stable, means (contributions of elements)? The first question is relatively easy to answer: Of course, unstable actions (movements, percepts, messages, and thoughts) are next to useless in most everyday situations because of the changing external conditions and intrinsic states of the living system itself. For example, to catch a prey or to avoid a predator, one better has a very stable, veridical picture of the environment with the prey/predator moving in it and a mechanism allowing stable, reliable actions in the environment despite possible unpredictable factors such as wind, stepping on a pebble, seeing an unexpected object, etc. The second question is much less obvious. Indeed, why is it advantageous being noisy at the level of elements? Further, I will try to offer an answer to this question and to merge the notions of activity and stability into a single coherent scheme that allows distinguishing behaviors of living systems and inanimate systems.

Definition of living system

2

Within this book, I accept the physical approach to living systems. This means that the ultimate goal of this line of inquiry is to discover *laws of nature* that define processes within living systems and interactions between living systems and their environments (which may include other living systems). This goal is different, for example, from trying to design and build an artificial device (a robot or a prosthesis) that is able to reproduce certain salient characteristics of certain behaviors of a living system. I assume that living systems emerged and developed as a result of the natural evolutionary process, not a divine intervention or some other act of creation, for example, by aliens. As a result, these systems are expected to obey laws of nature, which may be defined as compact representations of our experiences, typically in the form of equations. The scientific approach focused on laws of nature is commonly addressed as *physics*. Accepting the physical approach is equivalent to creating an area of natural science dealing with living systems, *Physics of Living Systems*.

This term is not new. In fact, the author graduated many years ago in a department called "*Physics of Living Systems*" from the Moscow Institute of Physics and Technology, also known informally as *Fiztekh*. As I understand now, the name of that department had little to do with reality—physics of living systems did not exist at the time and hardly exists now—but rather represented a promise that maybe, at some time in future, alumni of that program would help to create this field of science.

The purpose of classical physics has been to unite our experiences into a relatively small set of basic laws. These laws are typically formulated as equations that link salient variables, which describe states of the systems of interest, and parameters. One may say that the variables are constrained by the laws of nature while parameters are not. For example, consider arguably the most famous law of nature, Newton's Second Law, $\mathbf{F} = m\mathbf{a}$. This equation links *force vector* ($\mathbf{F}$) acting on an object and *acceleration* ($\mathbf{a}$) of the object. These two variables are constrained by the law: If one of them changes, the other has to change as well. The parameter m (mass) describes specific objects and is not constrained by this law: The law is equally applicable to objects with large and small mass, and mass does not have to change with changes in force or acceleration.

Laws of classical physics are applicable to particular types of objects. For example, the mentioned Newton's Second Law is applicable to objects with inertia. The well-known Hooke's Law, $\Delta\mathbf{F} = -k\Delta\mathbf{x}$, is applicable to certain classes of deformable objects that resist deformation and accumulate potential energy, commonly addressed as *springs*. In this equation, deformation of an object ($\Delta\mathbf{x}$) is linked to a change in force ($\Delta\mathbf{F}$) between this object and the environment with the help of a parameter called *stiffness* ($k > 0$). Application of this equation to different classes of objects that do not deform but move (for example, animal joints and limbs) has led to much confusion

Physics of Biological Action and Perception. https://doi.org/10.1016/B978-0-12-819284-9.00002-4

in the field of biomechanics (reviewed in Latash and Zatsiorsky, 1993, 2016), including reports of negative stiffness, which makes no sense in classical mechanics.

Living systems are not expected to violate Fundamental Physical Laws (under this term I understand all the laws found in physics textbooks). These laws impose constraints on biological objects, but these objects are not driven by those laws. In his recent book, Misha Gromov (2018) states that compatibility with laws of physics (the author implied Fundamental Physical Laws) is only a miniscule part of the foundation for biology. Gromov cites Erwin Schrödinger, who in his classic "What Is Life?" (Schrödinger, 1944, 2012) suggested that living objects were likely to obey unknown to us laws of physics, which, after they have been discovered, would form an inherent part of physics in general. Along similar lines, Israel Gelfand stated that the problem was not in applying known to us mathematics to biology but in developing new biological chapters of mathematics (quoted in Latash, 2008).

It is commonly assumed in classical physics that, during typical times of observation, parameters of systems of interest remain unchanged or change at much slower rates compared with law-constrained variables. I am going to suggest that living systems differ from inanimate ones in a major way: *Living systems induce changes in their states by changing parameters of relevant biological laws of nature*. These are going to be addressed as *Biological Physical Laws* in contrast to the Fundamental Physical Laws, common for all objects, living, and inanimate. Here comes a simple, brief definition, which is a step toward development of *Physics of Living Systems*:

A living system is a system able to (1) unite Fundamental Physical Laws into chains and clusters leading to new stable and pervasive relations among physical variables and involving new parameters and (2) modify these parameters to perform actions. In other words, Biological Physical Laws emerge in living systems, and then these systems modify parameters of the new laws to induce behaviors.

This definition is applicable to living systems of different complexities, from a macromolecule to the world economy. Accepting this definition leads to a number of questions, and I would like to admit upfront that many of those are *beyond my comprehension* (BMC) at this time. For example, what are the processes that lead to changes in assumed parameters of Biological Physical Laws? In other words, why would an animal suddenly decide to run uphill? There has to be a purpose for this action. What are the processes leading to goals and purposes? What is the origin of Biological Physical Laws (assuming that they exist)? How do Biological Physical Laws emerge within an organism?

In this context, I am using the word "purpose" as a shortcut for processes with unknown physics (including physiology) that makes actions of a living system look meaningful and functional to an external observer. A detailed review of philosophical aspects related to teleology of processes in living systems can be found in McLaughlin (2001). There have been several general approaches to purposefulness of actions (*BMC*) including the following.

It is very likely that the processes of evolution and natural selection played an important role in the emergence and shaping of Biological Physical Laws, particularly those common across organisms and species (Brandon, 1990, 1997). The explanations based on evolution, however, are rather general and do not address the form of specific

Biological Physical Laws and their parameters. In particular, the laws underlying the evolutionary process have been described with claims that these laws have "no need to turn to physics" (McShea and Brandon, 2010, p. 110). As the main stated goal of the line of thinking advocated in this book is to move toward *physics* of living systems, evolutionary laws are clearly insufficient. For example, they cannot address Biological Physical Laws specific to an individual organism that may be developed in the process of skill acquisition. Note that some of the special skills (e.g., those involved in circus performance, playing musical instruments, and professional sports) may carry no obvious evolutionary advantage and in fact have negative impact on health, life expectancy, and chances to have healthy offspring.

Mathematical modeling approaches, including optimization methods, have been invoked as ways to describe purposeful actions (for reviews, see Prilutsky and Zatsiorsky, 2002; Todorov, 2004; Diedrichsen et al., 2010). Typically, such approaches assume that biological systems are lazy and frugal, even stingy: These approaches try to minimize a cost function related to a behavioral goal, compute time patterns of salient performance variables that lead to the minimal cost of the action, and compare them to observed behaviors. A close match between the computed and observed variables is viewed as evidence that the assumed criterion adequately reflects relevant biological processes. While these approaches may suggest important insights, they do not consider laws of nature that might lead to the emergence of goal-directed biological actions. I would like to emphasize an important difference between "being able to predict" and "knowing laws of nature leading to" a phenomenon; for example, a regression model may be well-suited for the former, but not for the latter.

The physical approach does not have room for computations within the system of interest. Computational tools may and should be used by researchers to describe laws of nature and relevant processes. However, assuming that the system of interest performs mathematical operations and then acts to achieve a precomputed result is equivalent to giving up attempts at understanding the relevant laws of nature. Imagine that Isaac Newton assumed that some structures on the Sun and planets somehow measured the distances to each other, got information on the masses, and computed forces that had to be applied to celestial bodies to produce their trajectories. This would not be hard to do given that the Kepler Laws had been known at the time, but such an approach would not move us toward understanding the law of gravity and the concept of gravitational field.

State variables that describe neural and motor elements involved in any action change over time. This allows describing biological actions as outputs of a dynamical system (for review, see Kugler and Turvey, 1987; Kelso, 1995). To control behavior of such a system, one has to change its parameters (Glansdorff and Prigogine, 1971). Consider a simple example of a mechanical pendulum consisting of a weightless rigid rod and a point mass at its end (Fig. 2.1). Its salient parameters include the length of the rod (L), the vector of gravity (**g**), and the coordinates of the point of suspension $\{x_0, y_0, z_0\}$. If you want to change motion of the pendulum, you would have to change these parameters: Changing the rod length will affect the period of motion, while changing the point of suspension will change the spatial coordinates of the motion. Of course, one can build an artificial system that would imitate motion of the pendulum with

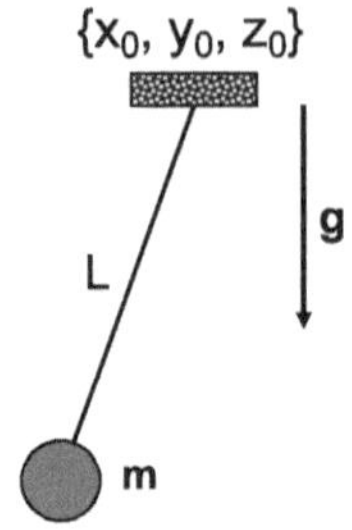

Figure 2.1 A mechanical pendulum in the gravity field (**g**) consisting of a weightless rigid rod with the length L and a point mass (m) at its end. The pendulum is suspended at a point in space with the coordinates $\{x_0, y_0, z_0\}$. To change motion of the pendulum, one has to change salient parameters.

high precision by placing a force-generating device (an actuator) between the rod and the mass, computing requisite time-varying forces, and applying them to the mass. This would allow producing any motion of the mass, including that typical of pendulum motion. But this exercise would only distract from understanding the physics of actual pendulum motion. This is another example of the qualitative difference between discovering laws of nature in a system and using computational means to reproduce the system's behavior. While behaviors of living systems can be well-approximated using computational methods borrowed from such fields as engineering and control theory, these methods are based on preprogramming of performance variables, which limits their usefulness for understanding the physics of dynamical (time-varying) systems involved in the production of biological movements.

To answer some of the commonly heard criticisms of this negative attitude to engineering (computational) approaches to biological problems, let me make a few statements. First, the fact that humans can perform computations does not mean that this is their innate ability and is irrelevant to everyday functions such as the ability to move and to perceive. One has to distinguish products of evolution and of civilization. Our basic functions, such as the ability to move and perceive, are products of evolution, and we may not be the best in some of those: Many animals move and perceive better than humans without any detectable ability to compute in their brains. We can compute because we were taught by teachers, not because our brain has evolved to possess computation as its inherent property. The last statement does not mean, of course, that in the process of evolution our genetic apparatus could not evolve in a way that gives us certain advantages with respect to acquisition of certain skills, both physical and mental. This could happen by chance (as mutations usually do), and then the society developed in directions where such skills could be used to their full potential. This was well-understood by some of Fiztekh professors who appreciated that computations were easy and not too imaginative. One of them (sorry, but I forgot the name!) used to say: "Solving equations is unseemly. You have to guess the answer and then prove that it is correct."

Second, humans are able to build nonliving systems that are able to show behaviors similar to those of animals. For example, cars can drive uphill and planes can fly against the wind. However, such objects do not emerge naturally, as a result of evolution, and they are typically limited in the available biologically looking behaviors. Finally, the engineering approaches to biological problems are very important for applied fields such as prosthetics and robotics. They are powerful if the goal is to build an artificial system that is able to emulate certain features of biological systems. They are of limited use, however, if the goal is to understand laws of nature that define behaviors of biological systems.

This seems to be perfect time to remind that many years ago Bernstein (1935) emphasized the impossibility to preprogram performance variables produced by biological systems in the changing environment. Therefore, from the physical perspective, *changing parameters is the only natural method of producing changes in performance variables during biological action*. Furthermore, I am going to address this method of control as *parametric control*.

2.1 Specific features of parametric control

To illustrate the specificity and advantages of parametric control, consider the following example. Imagine that you are playing billiards not on a typical rigid surface but on a horizontal elastic surface (Fig. 2.2). A motionless billiard ball would create on this surface a potential well (because of gravity) and rest on the bottom of this well in equilibrium. Imagine now that you want to move the ball to a new location, from point *a* to point *b* in Fig. 2.2. A traditional method would be to use a cue and apply force to the ball so that it rolls to the new desired location (shown in the middle drawing in Fig. 2.2). Computation of an adequate force time profile requires knowledge of the basic mechanical properties of the system such as mass of the ball, friction between the ball and the surface, and elastic properties of the membrane. If any of those characteristics are not known, most likely an error will take place and the ball will end up in a location different from the desired one. These problems can be solved if one has online information on ball motion (for example, with the help of vision)

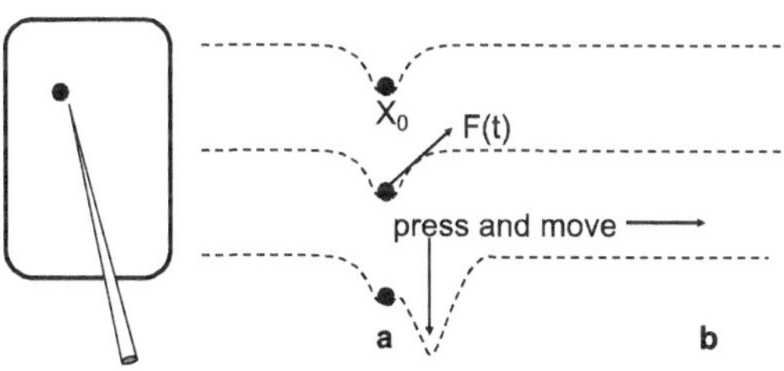

Figure 2.2 You are playing billiards on a horizontal elastic surface (left). A motionless billiard ball creates on this surface a potential well and rest on the bottom of this well in equilibrium (right top drawing). Imagine now that you want to move the ball from its current location (a) to a new one (b). One strategy is to compute and apply the requisite force profile (right middle drawing, **F**(t)). An alternative strategy is to modify the coordinate of the center of the potential well, X_0, for example by pressing with a finger and moving it from a to b (right bottom drawing).

and can quickly and adequately correct the force applied by the cue. This strategy would be an example of feedback control, which is used in numerous man-made machines.

Now consider how this task could be performed using parametric control. The potential well on the elastic membrane can be characterized by such parameters as its center coordinate (X_0) and elastic properties of the membrane that define the shape of the potential well. Imagine that you can change the former parameter. Then moving X_0 from the current location of the ball to its desired location would solve the problem. The ball would have no choice but to roll to the bottom of the potential well because no other equilibrium states were available. This method would solve the problem without an error (assuming that one can specify X_0 change with infinite precision, which is, of course, an exaggeration). Note also that this method does not require computation of forces to be applied to the ball. The forces will emerge as consequences of shifting X_0 and they will push the ball to the desired location.

How can this solution be realized? Actually, this is very easy. You should press with a finger on the elastic surface close to the ball (the bottom drawing in Fig. 2.2) and then, after the ball starts rolling toward the finger, move the finger to the desired location. Of course, finger motion should be produced at a reasonable speed so that the ball is not left too far behind where it could stop at an intermediate coordinate in a local potential well.

This example illustrates the main advantages of parametric control: (1) It does not require computation of mechanical variables such as forces; (2) it uses control variables (time changes in parameters of the system) at the level of task specification; and (3) it produces a stable solution. The first point does not require explanation. The second one means that the controller (the brain) does not have to compute any intermediate variables involved in biological movements, such as forces and muscle activations. It uses control variables that are direct reflections of task-specific variables. As most of actions by biological systems require motion to a spatial target, it makes sense to assume that control variables are also formulated as time changes in spatial coordinates (such as the time change of X_0 in the example of playing billiards on an elastic membrane). The third point means that this method of control can produce accurate solutions even if parameters of the moving object and external force field are not known precisely. For example, the method would work for balls of different mass and for different elastic properties of the horizontal surface, of course within reasonable limits. Note that if one uses the more traditional method of moving the ball, i.e., computing force and applying it to the ball, the solution becomes highly sensitive to the mentioned mechanical parameters. As a result, the trajectory and the final location are expected to show errors if some of the parameters happen to be different from those assumed at the step of force computation. In other words, the solution becomes unstable. It requires corrections in the process of motion to compensate for the errors made at the step of force computation.

Assuming that biological systems produce motion by changing parameters of relevant Biological Physical Laws raises one more issue. As these *parameters* are assumed to change at least as quickly as the induced changes in variables, should they not be addressed as *variables*? This sounds like a valid point, but only if one considers mathematics. In physics, parameters of natural laws differ from variables not only in their characteristic rates of change (to remind, typically, parameters change much slower than variables) but also, more importantly, by the fact that they are not constrained by the laws while variables are. Indeed, one can imagine a situation when mass of an object changes at a fast rate; for example, fuel in a rocket burns out quickly leading to a change in its mass. Nevertheless, the quickly changing mass remains a parameter in the Second Newton's Law, while force and acceleration remain variables.

While robotics is a field of engineering that seems to be very close to the neural control of biological movements, the main difference between artificial moving systems, including most sophisticated ones, and biological movements continues to persist. Robots are powered by actuators, i.e., motors that produce programmed forces and moments of force. Given the powerful motors and feedback loops with negligible time delays, robots are able to emulate many salient features of natural animal movements (see the movies of very impressive robots available on the Internet, e.g., those produced by *Boston Dynamics*). Nevertheless, the principle of control remains the same as in the inanimate nature: To produce movements, forces are precomputed and applied. Attempts at using parametric control in artificial systems remain few (cf. Bicchi et al., 2011: Godfrey et al., 2013).

Some robots can emulate stability features of natural animal (including human) movements. However, stability in such robots is not task-specific, it is part of the design of the system. For example, underactuated robots (reviewed in Birglen et al., 2008; Xin and Liu, 2014) use fewer actuators than kinematic degrees of freedom (typically, axes of joint rotation that can, at least hypothetically, move independently of each other). The remaining kinematic degrees of freedom are spanned by passive devices with viscoelastic properties, which naturally ensure local stabilizing action. For example, an artificial three-joint leg may be equipped with only two motors acting at the knee and hip joints, while the ankle joint may be spanned by a damped spring with constant parameters. Such devices show trial-to-trial (or step-to-step) variability patterns that remind those seen during animal locomotion, reflecting stability of the locomotion pattern. These will be described a bit later. However, even in such systems, the principle of control is based on specifying torques at the joints powered by actuators, although not at all the joints, and does not include parametric control, which distinguishes biological motion from inanimate motion.

2.2 Examples of Biological Physical Laws from movement science

2.2.1 Stretch reflex

Arguably, the best-known and most universal Biological Physical Law in the field of movement science is the dependence of active muscle force on muscle length. If a muscle, with its neural connections to the spinal cord intact, is slowly stretched by

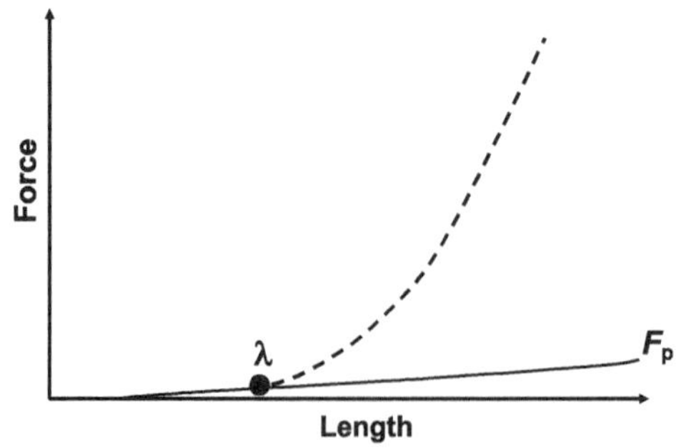

Figure 2.3 If a muscle is slowly stretched, it generates resistive force F even if it is quiescent (F_P, the *solid line*). At some magnitude of length L (λ, threshold of the stretch reflex), the muscle shows first signs of activation. This is accompanied by a steeper $F(L)$ dependence (the *dashed line* in Fig. 2.3).

an external force, it generates resistive force even if it is quiescent (shows no signs of electrical activation) because of material properties of peripheral tissues (see the F_P characteristic in Fig. 2.3). At some magnitude of length, the muscle shows first signs of activation, and its activation level increases with further stretch. This is accompanied by a steeper $F(L)$ dependence (the dashed line in Fig. 2.3) that deviates from the F_P characteristic. The change in activation is induced primarily by reflex effects from sensory endings in the muscle and surrounding tissues sensitive to muscle stretch including those in structures called *muscle spindles*. Altogether, these effects are addressed collectively as *stretch reflex* (Liddell and Sherrington, 1924). Therefore, the muscle length at which first signs of muscle activation are seen has been addressed as *threshold of the stretch reflex* (λ). The dependence of active muscle force (the excess of force over F_P) on muscle length can be described as

$$F_A = 0, \text{ when } L \leq \lambda \tag{2.1}$$

$$F_A = f(L - \lambda) \text{when } L > \lambda$$

Let us, for now, disregard the exact form of the function f. What matters is that two salient state variables for an intact muscle, F_A and L, are linked with the help of a parameter λ. According to one of the most influential hypotheses in motor control, the equilibrium-point hypothesis (Feldman, 1966, 1986), the central nervous system controls individual muscles by specifying time changes in corresponding λs (see in detail in Section 5.1).

This Biological Physical Law is based on many events within the body that may be viewed as chains and clusters of elementary Biological Physical Law, while each elementary Biological Physical Law is ultimately based on chains and clusters of

Fundamental Physical Laws. For example, consider the following sequence of events leading to the observable $F_A(L)$ relation:

(1) muscle stretch leads to deformation of muscle fibers and length-sensitive sensory endings located in muscle spindles—complex structures within skeletal muscles that house numerous sensory endings;
(2) the endings respond to deformation by changes in their membrane permeability to certain ions leading to the generation of action potentials—units of information transmission within the nervous system;
(3) the action potentials are transmitted along the axons of proprioceptive neurons into the central nervous system;
(4) they lead to release of neuroactive chemicals (neurotransmitters) at the target connections (synapses) with other neurons;
(5) this leads to changes in the membrane potential of the target neurons and the generation of action potentials by some of those neurons;
(6) these neurons make connections (synapses) with other neurons within the central nervous system;
(7) ultimately, some of the large output neurons, the alpha motoneurons innervating the muscle, generate action potentials;
(8) those action potentials are transmitted along the axons of the alpha-motoneurons and act at the neuromuscular synapses leading to release of the appropriate neurotransmitter;
(9) this leads to muscle membrane potential changes and action potential generation on the muscle fiber membrane; and
(10) after a complex excitation–contraction coupling process, including the release of calcium ions from their storage place within the muscle fibers and the formation of links between long macromolecules (cross-bridges), electrical signals on muscle fibers lead to the generation of force and muscle fiber shortening.

In fact, each of the mentioned ten steps is by itself a complex process based on more elementary components. However, we hope that, ultimately, each elementary step can be reduced to a combination of Fundamental Physical Laws.

Of course, the stretch reflex (as well as other examples in this paper) can be simulated using computational means. In fact, any law of physics, biology-specific or not, may be simulated by other means. For example, one can build a computational model of interactions between two charged bodies, without invoking the notion of electromagnetic field, to match the actual behavior of these bodies with high accuracy. However, such a model would not be equivalent to knowing the concept of electromagnetic field and the Coulomb Law that bring about the forces during interactions of charged bodies.

2.2.2 Other reflexes as Biological Physical Laws

There are other, well-established dependences that link state variables at different levels of description of the neuromotor system in ways that are not immediately obvious from Fundamental Physical Laws. Some of these dependences are relatively stereotypical and have been traditionally addressed as *reflexes*.

While there seems to be shared intuitive understanding what a reflex is, there is no unambiguous definition for reflex. Commonly, reflexes are considered in contrast to voluntary actions and the definitions of reflex emphasize its link to a stimulus, its involuntary nature, its stereotypical pattern, and its quickness. In particular, Wikipedia defines reflex as an involuntary and nearly instantaneous movement in response to a stimulus. This definition is not very useful because it uses the word *involuntary*, which is as undefined as *reflex* is. Besides, "nearly instantaneous" is also marginally helpful and may even be misleading for those who know basic physics because it seems to imply purely electrical mechanisms of signal transmission. Other definitions use related constructs such as "response not reaching the level of consciousness," "action consisting of simple segments of behavior," "unique correlation of response with the stimulus," etc.

While many of such definitions are intuitively appealing, they use other undefined or loosely defined notions and characteristics to define reflex. A recent paper co-authored by several prominent scientists (Prochazka et al., 2000) asked a question whether *reflex* was a useful notion or it was too ambiguous to be used in research on biological movement. The authors did not come to a consensus, but the discussion emphasized two main polar views: (1) *Reflex* is a misnomer, it implies an action that may look simple but otherwise not different from any other action including voluntary movements and (2) *reflex* is a useful notion, it differs in an important way from voluntary movement.

Fig. 2.4 contrasts reflexes and voluntary movements with respect to several characteristics that have been most frequently mentioned among the definitions of reflex. Note that the figure implies a continuous spectrum along each of the axes.

Consider the first characteristic, link to stimulus, with respect to the notion of stretch reflex, which is one of the best-established reflexes. Its origin is primarily the activity of length-sensitive receptors in the muscles, namely sensitive endings in muscle spindles. When someone taps your patellar tendon, a brisk knee extension movement is observed at a very short delay (about 25–30 ms), which is traditionally addressed as a *reflex* produced by an abrupt activation of the muscle spindle sensory endings. However, during natural everyday movements, muscle spindle receptors are active most of the time. As a result, there is no clear identifiable stimulus that leads to stretch reflex unless one views each and every action potential generated by spindles as

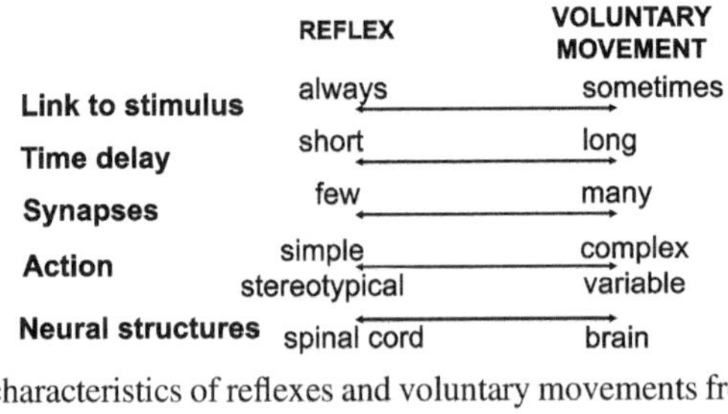

Figure 2.4 Several characteristics of reflexes and voluntary movements frequently mentioned in the definitions of reflex.

a separate stimulus. Therefore, it is not easy to link the dependence of muscle force on its length illustrated in Fig. 2.3 to a stimulus. On the other hand, many everyday voluntary actions are performed in response to a stimulus. For example, seeing a red light causes the driver to press the brake pedal. In laboratories, participants are commonly asked to press a button (or perform another simple action) as quickly as they can after they see or hear a specific stimulus. Commonly, people who perform such actions very quickly and reliably are described as persons "with good reflexes." This is jargon, of course, because the action they perform is definitely a voluntary action: In all the mentioned cases, the actor could ignore the situation on the road or the instruction from the experimenter and produce no action at all.

Other characteristics are similarly questionable. For example, with respect to the time delay, muscle activation in response to high-frequency vibration applied to the muscle or to its tendon (the tonic vibration reflex, Eklund and Hagbarth, 1966; see also Section 6.3) typically takes relatively long time, commonly over 1 s, to show discernible muscle activation. This time delay may be much longer than typical times of initiating a quick voluntary action to a stimulus.

Other criteria also cannot distinguish reflexes from voluntary actions in a reliable way. There are reflexes involving numerous synapses, some reflexes lead to complex coordinated actions, some reflex responses are highly variable, and some reflexes are mediated by supraspinal structures, those within the brain stem and even involving the cortex of the large hemispheres of the brain.

Despite the lack of a universal, unambiguous definition, it seems fair to say that reflexes link sensory signals produced by some physical variables (such as forces, displacements, luminance, sound waves, etc.) to changes in muscle activations resulting in action, which can also be described as a change in certain physical variables. For example, the pupillary reflex leads to a change in the size of the pupil caused by changes in the overall level of luminance. This reflex involves a complex neural loop originating from neurons in the retina and involving the pretectal nucleus and the Edinger–Westphal nucleus in the midbrain, the ciliary ganglion, and the ciliary muscle. Its action may be described formally as a function that links the pupil diameter and the level of luminance, i.e., a Biological Physical Law.

Another example is the wiping reflex that can be seen in some animals, e.g., in frogs, with the spinal cord surgically separated from the brain (Fukson et al., 1980; Berkinblit et al., 1986). In these animals, movements of the trunk and the legs are controlled exclusively by the spinal cord and are commonly considered as reflexes. The wiping reflex is triggered by an irritating stimulus (e.g., a small piece of paper soaked in a weak acid solution) placed on the frog's body. It leads to movements in multiple joints, which result in a complex movement of the hindlimb on the same side of the body. The movement brings the toes close to the stimulus and then wipes if off the body with a quick extension in all the major limb joints. This reflex action is highly variable and shows amazing versatility. In particular, the frog can produce accurate wiping action when the hindlimb is loaded with a lead bracelet or when one of the joints of the hindlimb is blocked from moving by an orthotic device (Latash, 1993, 2008).

2.2.3 Predictive (feedforward) aspects of behaviors

Animals frequently behave in a predictive fashion. For example, a predator typically runs not to the place where it sees a moving prey but to a place where it expects to intercept the prey. Two approaches to such actions or action components have been dominating the field of motor control. One of them considers such feedforward actions or adjustments to actions as results of a computational process with a few salient variables. Consider the following two examples. Both examples describe phenomena that are seen across populations and tasks; they are "automatic" in a sense that actors are typically unaware of these adjustments of actions.

It has been well-known that humans change the grip force applied to a handheld object in a feedforward fashion before manipulating this object (reviewed in Zatsiorsky and Latash, 2008). A traditional, teleological explanation of these phenomena has been formulated as trying to keep the safety margin (extra gripping force over the threshold for slippage) at a subject-selected constant magnitude (Johansson and Westling, 1984). In other words, the subject's brain estimates the expected force tangential to the grasped surface (Fig. 2.5), uses the known (estimated) friction coefficient between the digits and the surface of the object, computes the required grasping force to keep safety margin at a certain level, and sends signals to the spinal cord and muscles that are adequate to generate the required forces. This account makes a few assumptions, the strongest one arguably being that the central nervous system performs computations and is able to program forces exerted by the effectors on the environment.

Another example is the phenomena of anticipatory postural adjustments (APAs) before self-triggered or externally triggered but predictable postural perturbations (Belenkiy et al., 1967; Massion, 1992). These adjustments involve changes in the activation of muscles involved in the postural task component seen about 100 ± 50 ms before first signs of the perturbation. For example, when a standing person performs a very fast arm movement, e.g., extends the arms, first signs of changes in background muscle activation are seen not in the prime movers for the task (e.g., the anterior deltoid muscle) but in muscles of the legs and trunk. Quantitative characteristics of these adjustments depend on a number of factors

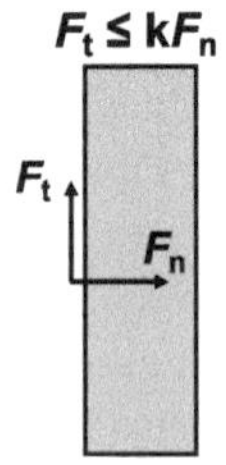

Figure 2.5 If one wants to move a vertically oriented object grasped with two digits upwards, the grip force normal to the surface (F_n, shown for one digit only) changes in anticipation of the changes in the tangential force (F_t), which is needed to move the object. This allows satisfying the inequality $F_t \leq kF_n$ and avoiding slippage (k is the friction coefficient).

such as the site and magnitude of perturbation, stability of posture, and magnitude of action associated with the perturbation (reviewed in Aruin, 2003). APAs are also observed before expected perturbations not associated with overt actions, e.g., during preparation to an impact from an approaching object, for example, during catching an object. A traditional interpretation of APAs has been that the central nervous system estimates forces and moments expected from the planned action and then generates forces and moments in opposite direction so that the net effects of the perturbation are reduced or even canceled out. I would like to emphasize that this interpretation is also based on assuming that the central nervous system performs computational operations and programs muscle activation adequate for the generation of requisite forces and moments.

As we will see in one of the further sections, these assumptions are likely false or, at least, misleading. None of the mentioned interpretations tries to describe laws of nature leading to such feedforward adjustments. An important step in this direction has been made recently by Stepp and Turvey (2010, 2017) who emphasize the difference between so-called *weak anticipation* and *strong anticipation*: Weak anticipation is based on estimations and computations, while strong anticipation is a consequence of laws of nature.

2.2.4 Where do Biological Physical Laws come from?

Some of the aforementioned Biological Physical Laws are consequences of evolution. For example, all the laws common across species, such as the stretch reflex, have probably developed relatively early in the evolutionary process. These Laws require massive individual experience to lead to a visible modification, and the magnitude of such a modification is relatively modest. For example, in their classical studies, Jonathan Wolpaw and colleagues (Wolpaw and Carp, 1993; Wolpaw and Tennisson, 2001; Wolpaw, 2007) explored whether individual experience could produce a persistent change in the magnitude of the monosynaptic reflexes induced in limb muscles of a monkey. The monosynaptic reflexes are the simplest reflex loops that involve only one connection between neural cells (a neuro-neural synapse) in the spinal cord, from a proprioceptive neuron equipped with a sensory ending naturally sensitive to muscle length and its velocity and an alpha-motoneuron innervating the muscle. The magnitude of this reflex shows natural trial-to-trial variability. The monkeys were given reward when the reflex amplitude exceeded a certain criterion value. Therefore, to optimize the reward amount, the monkeys were expected to find a way to produce consistently higher reflexes in that muscle. This is a typical operant conditioning design, when an animal is not shown a desired solution for a task but is expected to discover such a solution by trial and error. In these studies, it took thousands of repetitions over multiple sessions to produce a persistent increase (or decrease) in the monosynaptic reflexes. On the one hand, the study carried an important message that even the simplest reflexes could be modified with operant conditioning techniques. On the other hand, it showed that producing such an effect in spinal reflexes takes an enormous effort and time on the part of the animal.

Several later studies tried to use a similar technique to modify monosynaptic reflexes in humans, in particular in patients suffering from spasticity, a condition with exaggerated reflexes and muscle spasms (reviewed in Nielsen et al., 2007; Dietz and Sinkjaer, 2012; see also Section 5.7.2). Those studies reported much quicker changes in the relative magnitude of the monosynaptic reflexes in the limb muscles, typically a drop in the reflex magnitude, which were seen after only a few dozen of repetitions (Thompson and Wolpaw, 2015; Nakajima et al., 2016). It is possible, however, that the subjects were helped by "cheat strategies"; for example, they could discover, likely without a conscious effort, that muscle contractions in other parts of the body, which might not be controlled in the experiment, could lead to the desired reflex change. Note that it has been known for many years that contracting remote muscles may have a significant effect on the magnitude of monosynaptic reflexes in the legs. The best-know example is arguably the Jendrassik maneuver—a strong contraction of arm muscles or clenching the teeth—which leads to large changes in the leg monosynaptic reflexes (Delwaide and Toulouse, 1983; Gregory et al., 2001), possibly mediated by propriospinal neural pathway connecting different segments of the spinal cord. It is also possible that the humans, but not monkeys, could readily conceptualize the induced reflex amplitude as the goal of the task and then use motor imagery to facilitate or attenuate the responses in the target muscle group; note that effects of motor imagery on reflex amplitude have been demonstrated (Aoyama and Kaneko, 2011; Traverse et al., 2018).

Biological Physical Laws developed in the process of evolution are expected to give an advantage to their carriers. For example, most spinal reflexes shared by a large number of species are negative feedback loops that help stabilize posture of and action by the effectors.

Another method of developing a Biological Physical Law is building a motor skill, for example, as a result of practice. Consider, for example, the skill of riding a bicycle. Developing this skill requires suppressing the natural tendency to move away from the direction of fall as a means of recovering vertical posture. There are two basic methods of recovering vertical orientation when the center of mass happens to project outside the support area. The first is to apply forces to the environment that would create reaction forces moving projection of the center of mass toward the support area. The second is to move the support area toward the current location of the center of mass projection. The first method is used when the supporting surface is stable and the second one when it is movable. Examples of the latter include walking on the tightrope and riding the bicycle. This new skill—turning the handles toward the fall direction—is learned relatively quickly and without conscious understanding. Furthermore, it becomes so strong that unlearning it turns into a nearly insurmountable problem.

This problem has been illustrated by a web-based clip of a "crooked bicycle," which inverted the relation between the handle motion and front wheel turn. In that bicycle, to turn the front wheel to the right, the handles had to be turned to the left. The author of this contraption told that it had taken him several months to learn how to ride this bicycle without falling down. From the point of view of neural computations, the task is trivial: multiplying one of the variables in the previously learned rule by (−1). The fact that learning this new skill required months of practice shows that skill formation is not based on neural computations. After the person learned how to ride the "crooked bicycle," he lost completely the preexistent ability to ride a "normal bicycle." The two skills were seemingly incompatible! After a skill is learned, it requires no conscious effort and becomes a reflection of the corresponding new Biological Physical Law involving a few parameters.

The last examples suggest that at least some of the laws of nature are not eternal and invariable but can change in the process of biological evolution and even within the lifetime of a single organism. This point is directly related to the classical discussion between two great thinkers, Albert Einstein (1879−1955) and Henri-Louis Bergson (1859−1941), on the mathematical expression of laws of nature (reviewed in Canales, 2015). The first view follows the traditions of classical physics that fundamental laws of nature are eternal, not changing with time, whereas time is just another coordinate in the four-dimensional space-time World. Einstein accepted this tradition and viewed the ability to express laws of nature in a compact (and elegant!) mathematical form, a miracle. The alternative view is that the World changes with time in the continuous process of evolution and these changes can be accompanied by changes in laws of nature. These changes make immediate experience more valuable than the abstract, rationale scientific approach. Within this line of thinking, time is not simply a fourth coordinate but special in its ability to bring about changes in laws of nature, at least those that are relevant to objects evolving with time. Biological objects definitely belong to that class.

Redundancy and abundance

3

We are now approaching a group of problems that have been in the center of attention of the neuroscience research community for nearly one hundred years. Similar problems are encountered in other areas, such as engineering and control theory, reflecting an increase in the complexity of machines designed by engineers that require more and more sophisticated means of control. The success of engineering, in particular in the field of robotics, encouraged researchers working with biological systems to borrow tools from those fields and apply them to biological problems. The first results were encouraging: The tools worked relatively well when applied to simple, constrained actions. However, although more natural biological actions seemed not to violate any of the engineering principles by too much, they have also refused to be driven by those principles.

Many years ago, Israel Gelfand stated that any field of science had to develop a small set of adequate notions (an adequate language; Gelfand, 1991; Gelfand and Latash, 1998) that would separate it from other fields of science. He used the examples of the notions of *point* in the development of geometry and of *force* in the development of physics. (Indeed, Euclid defined point as an object with no height, width, and length. It is obvious that such objects do not exist. However, this brave step was crucial in defining line, plane, and other notions indispensable for geometry.) Gelfand suggested that, to become a successful field of science, biology had to develop its own adequate language, not to borrow languages from other fields, even if those fields were very well developed. This does not mean, of course, that biology should ignore progress in other fields (for example, classical mechanics is one of the pillars of human movement science) but that researchers should not expect other fields of science to solve biological problems.

The fields of engineering and control theory were developed by humans to deal with man-made machines, not with living systems that have emerged in the process of evolution. Unless we assume that animal (including human) bodies were designed by a Supreme Engineer, there is no reason to expect that engineering would help to discover laws that lead to natural behavior of animals. In other words, application of engineering to biological problems is an example of lightly veiled creationism.

This is an important point, and let me restate it in more general terms. If a biological system seems to have a disadvantage when compared with a man-made device, the first question should not be: How can one fix the suboptimal biological design to make it behave comparable with the man-made machine? or What kind of a computational mechanism should be placed into the brain to handle the apparent deficiency of the biological design? The first question should be: What evolutionary advantage is afforded by the design of the biological system?

Physics of Biological Action and Perception. https://doi.org/10.1016/B978-0-12-819284-9.00003-6

Consider the following examples. Human muscles are much slower and less predictable in their behavior compared to torque motors. Compared to artificial sensors, sensory endings in the human body produce noisy and ambiguous information with respect to salient variables such as forces and coordinates. Signals within the central nervous system are transmitted at a relatively low speed compared with the speed of electrical signals in artificial control circuits leading to substantial time delays. All these features seem to make the design of the human body suboptimal compared to artificial systems, such as robots. And indeed, given a narrow range of tasks, e.g., lifting heavy objects, moving at a high speed, and producing accurate manipulations in perfectly predictable environments, machines easily outperform humans. Nevertheless, robots are still very far from matching humans in more natural tasks that have to be performed in the changing and unpredictable environment.

Long time ago, in the Soviet Union, one common way to retain sanity under the crazy communist regime was telling jokes. Here is one of them related to the previous couple of paragraphs. A military general shares a drink with his high school friend who works as a university professor: "I know that you, the academics, tell jokes about us, the military. All your jokes are about generals being very stupid and academics being very smart. But if you are so smart, why don't you march along the streets in regiments"?

In this section, we will first consider central problems inherent to numerous functions of the human body from the point of view of engineering and control theory. As the former paragraphs show, the author does not care too much about those approaches, but they follow rather directly the original formulation of those problems by Nikolai Bernstein. Besides, one should know the history of science to avoid repeating mistakes and to appreciate alternative ideas.

3.1 Problems of motor redundancy

One of the features of the human body is the apparent excess of elements at any level of analysis of any of its numerous functions. For example, in the motor domain, a person can hold an object in the hand with an infinite number of combinations of forces and moments produced by the individual digits; it is possible to bring the index fingertip to a spatial target with an infinite number of combinations of individual joint rotations; one can produce torque in a typical joint with an infinite number of combinations of forces produced by the individual muscles crossing the joint, etc. In the perceptual domain, we can use information from different modalities to perceive location of a body part in external space, visual, somatosensory, and in some conditions even auditory. In the language domain, a thought can be expressed using numerous grammatically correct phrases (assuming that the thought is simple enough to be expressed adequately using human language—see Section 7.1).

Consider the following problem: How can the tip of the index finger be brought to a target in space (Fig. 3.1)? The target can be described with three numbers, or coordinates, in the external space with respect to the body (assume that the target is close to

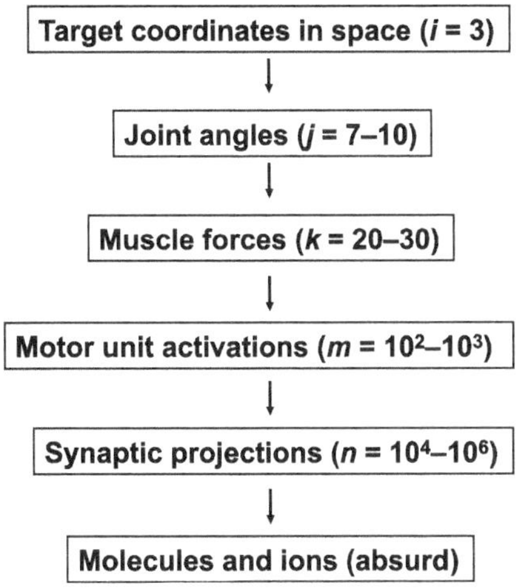

Figure 3.1 Abundant transformations involved in a task of moving a hand to a spatial target. Note that $n > m > k > j > i$.

the body so that the trunk does not have to move). To move the finger, one has to produce rotations in many joints that link the fingertip to the trunk: There are at least seven individual axes of rotation in the major arm joints (or ten if one considers scapular motion), in addition to possible motion of the within-the-hand joints. The problem of defining rotation in 7 or 10 joints to bring the fingertip to a three-dimensional spatial target belongs to the class of ill-posed problems. It is similar to solving three equations with 7 (or 10) unknowns. This cannot be done unless one comes up with more equations or other means of finding a unique solution. Nikolai Bernstein was the first to bring this problem into the focus of attention of the scientific community (Bernstein, 1947, 1967). Hence, it is commonly referred to as the *Bernstein problem* (Turvey, 1990) or the *problem of motor redundancy*. Bernstein formulated this problem as the one of *elimination of redundant degrees of freedom*. In the author's subjective opinion, this formulation and the use of the loaded term *redundancy* slowed down progress in the field of motor control for at least half a century. So, if one quantifies greatness of a scientist with the number of years he or she slowed down progress in science (see Resnick, 2015), Bernstein definitely was a great scientist.

The problem does not stop at the level of joint rotations, of course. Imagine that joint rotations are somehow defined. Joints are moved by changes in joint torques that are consequences of changes in muscle forces. As typical joints are crossed by several muscles, the problem of finding muscle force patterns that would produce a desired joint rotation is also ill-posed. For example, the elbow joint with only one kinematic degree of freedom (ignoring, for the sake of discussion, its involvement into pronation–supination) is crossed by three flexors and three extensors. How can one define forces of the six individual muscles to match the desired torque magnitude or time profile? This is equivalent to solving one equation with six unknowns—another ill-posed problem.

Let us skip the problem of finding muscle activations that would produce desired muscle forces (by itself, very non-trivial!). Muscles are composed of many motor units, groups of muscle fibers controlled by a single neuron, an alpha-motoneuron, located in the spinal cord or in the brain stem (for the neck and facial muscles). Individual motor units can be recruited at different frequencies of the generation of action potentials, which change their contribution to total muscle activation. So, the problem of defining a pattern of motor unit recruitment to match a desired level of overall muscle activation is also ill-posed, equivalent to solving one equation with potentially hundreds of unknowns.

One can continue and ask the question: How can one define individual inputs to an alpha motoneuron based on its desired output? There are many thousands of inputs to a neuron from other neurons, and the output of the target neuron is defined by the net effect of these inputs on the membrane potential. This problem is very very very redundant. (Later, we will also see that this problem is not solvable because of the threshold properties of the neuronal membrane.) And finally, to emphasize the absurdity of this logic, let us ask the question: Which of the ions should cross the membrane to produce a single output (an action potential) of a neuron? There are zillions of sodium and potassium ions in the vicinity of the membrane, and a few hundred thousand of those should cross the membrane to produce an action potential. Which ones? This is an obviously ridiculous question because the organism does not care which particular ions cross the membrane.

But does the organism care about all other ill-posed problems? Does it care about patterns of motor unit recruitment, muscle force combinations, and individual joint rotations? Or does it only care about the ultimate goal of the action? Different researchers would probably give you different answers to these questions.

There is a related problem of selecting a particular trajectory of a variable from its initial value to a desired final value, even if only a single variable is involved. For example, if one wants to move a joint from 30 degree to 100 degree within 1 s, there is an infinite number of joint trajectories solving the problem. For example, one can start the movement very fast and then approach slowly the final position or start the movement very slowly and then move very fast to the final position. This problem is applicable to both one-element and multielement movements. To distinguish between the two problems, we are going to address the classical Bernstein problem caused by excess of elements as the *problem of state redundancy*. The problem of selecting a specific trajectory will be addressed as the *problem of trajectory redundancy*. Commonly, both problems coexist as natural movements involve multiple elements and allow freedom in reaching a target state via various trajectories.

3.1.1 Optimization of movements

As mentioned in Part 2, a broadly used method to problems of motor control is that of optimization based on the principle of economy applied to a system involving a redundant set of elemental variables. In such systems, motor tasks typically have an infinite number of solutions, and a single solution is selected based on an *a priori* criterion. Such criteria are formulated as requirements to satisfy the task constraints and simultaneously minimize (or, less frequently, maximize) value of a *cost function* (also addressed as an *objective function*). Although the range of cost functions used in the field of motor control is rather broad, they all assume that the central nervous system tries to spend as little as possible of a resource that it views as potentially valuable and limiting performance.

A common general formulation of the problem of optimization is in the form of *minimization of the norm* of the solution for an action involving a change in a set of elemental variables (ΔEV). Formally, this means minimizing the sum $\sum \Delta EV^n$, where n is the power. Assuming that changing an elemental variable is associated with a certain cost, this criterion minimizes total cost associated with changing all the elemental variables. For $n = 2$, this criterion is equivalent to minimizing the integral of the Eucledian distance between successive points along a trajectory connecting the initial and final states in the elemental variable space, which is addressed in robotics as *Moore–Penrose* pseudo-inverse (e.g., Whitney, 1969).

Motion of any redundant system can be viewed as consisting of two components, within the solution space and orthogonal to that space. The direction defined using the minimum-norm criterion (e.g., using the Moore–Penrose pseudo-inverse) is orthogonal to the solution space and is commonly addressed as *range–space motion*. Movement within the solution space is addressed as *self-motion*. By definition, self-motion of a system leads to changes in the values of the elemental variables without an effect on the task-related performance variable to which the elemental variables contribute. The main assumption of the minimum-norm approaches is that the controller tries to minimize self-motion, which seems a reasonable assumption, which can be seen as a strategy to avoid wasting energy.

Although humans sometimes produce trajectories that obey the minimum-norm criterion (e.g., in multifinger tasks, Zatsiorsky et al., 2002), other studies have provided evidence for large amounts of self-motion. In particular, large amounts of self-motion were observed in experiments with unexpected perturbations applied during multijoint-reaching tasks analyzed in spaces of joint angles (Mattos et al., 2011) and muscle activations (Mattos et al., 2013) and also in multifinger tasks analyzed in spaces of finger forces and finger modes—hypothetical control variables to individual fingers (Mattos et al., 2015). Taken together, these results suggest that while the minimum-norm criterion may be able to account for some behaviors, its area of applicability is likely limited.

A number of cost functions based on minimization of mechanical variables have been suggested to describe movements to a target in conditions of trajectory redundancy. Some of them have been based on kinematic variables. These include the *minimum-time* criterion (Nelson 1983), which involves applying maximal values of acceleration and deceleration with a switch between the two at an optimal time. The optimal switch time may depend on a number of factors such as friction, relative power of the movement agonists and antagonists, etc. For a purely inertial system with equally powerful agonist and antagonist muscle groups, the optimal switching time is exactly in the midpoint of the movement. This criterion, however, produces unrealistic time profiles of kinematic variables including instantaneous changes in acceleration.

Arguably the most commonly used cost function in studies of movement kinematics, the *minimum-jerk* criterion, produced bell-shaped velocities resembling those seen during natural movement (Hogan, 1984; Flash and Hogan, 1985). According to this criterion, the optimal trajectory is the one minimizing the integral over movement time of the third time derivative of the coordinate (jerk, $J = d^3x/dt^3$) squared: $J = \frac{1}{MT}\int_0^{MT}\left(d^3x/dt^3\right)^2 dt$, where MT stands for movement time. This criterion, as well as many other optimization approaches, requires knowing movement time in advance, which makes it unlikely to be used in planning or executing natural movements when movement time cannot be known precisely before movement initiation.

Other cost functions have been formulated in terms of forces and force-based computations based on mechanical models of the moving effectors. In particular, one of the approaches tried to extend the minimum-jerk criterion to kinetic variables. This attempt resulted in the *minimum-torque change* model (Uno et al., 1989): $C = \frac{1}{MT}\int_0^{MT}(dT/dt)^2 dt$, where C is the cost function and T is torque. For a purely inertial one-element system, this criterion produces the same trajectories as the minimum-jerk one. The two approaches differ for more realistic models of moving systems with mechanical properties not limited to inertia. In such cases, the minimum-jerk criterion produces more realistic predictions.

Other cost functions are based on more complex variables that try to address physiologically important processes, which accompany actions, such as energy expenditure and fatigue. The *minimum-fatigue* criterion was based on experimental observations that linked endurance time, defined as the time during which a muscle can produce a certain force level, to muscle stress, that is, the ratio of force to the physiological cross-sectional area of muscle fibers (Crowninshield and Brand, 1981). The inverse of endurance time may be associated with muscle fatigue leading to the following cost function: $C = \sum_{i=1}^{n}(F_i/PCSA_i)^3$, where n stands for the number of muscles.

Estimating metabolic costs of an action is not easy, given that muscles spend energy to perform negative work, i.e. when they are involved in an action associated with muscle stretch (Latash and Zatsiorsky, 2016). A *minimum–metabolic cost* criterion (Alexander, 2002) was formulated as $C = \sum_{i=1}^{n}\alpha_i^p F_i^{\max} V_i^{\max}\phi\left(V_i/V_i^{\max}\right)$, where function ϕ determines the metabolic cost, $F_i^{\max}$ and $V_i^{\max}$ are known maximum force and maximum velocity of each of the n muscles, respectively, V_i is instantaneous velocity of the ith muscle, and α_i ($0 \le \alpha_i \le 1$) is the unknown normalized activation of the ith muscle, which is defined by minimizing the cost function C. Minimizing energy expenditure has an evident biological advantage, in particular, for actions that last for long times, such as locomotion. Indeed, it has been suggested that, at a given speed of locomotion, people and animals select the stride length and frequency at which the energy expenditure is minimal (Ralston and Lukin, 1969).

A cost function combining neural and mechanical variables was suggested based on the idea of control with shifts of equilibrium positions (Hasan, 1986; see Part 5).

This *minimum-effort* criterion minimized the integrated squared velocities of the equilibrium positions over movement time, taking into account a parameter representing joint apparent stiffness (see Section 5.6.4): $C = \int_0^{MT} k(d\beta/dt)^2 dt$, where β is equilibrium position, MT is movement time, and k stands for joint apparent stiffness. The application of this criterion allowed estimating optimal values of joint apparent stiffness; note that this criterion also requires knowing movement time in advance, which is not always compatible with natural actions.

Other cost functions involved variables related to notions from psychology that were difficult to measure in experiments and complex cost functions. One of the approaches considered *discomfort* as a variable minimized during movements (Cruse and Bruwer 1987; Cruse et al., 1990). Discomfort scores were associated with subjective reports by the subjects for different joint positions, which were integrated over all joints involved in movements. Discomfort functions typically showed low scores in the midrange and higher scores as the joint approached its limit of rotation. In the midrange, the scores changed very little with joint rotation and hence showed low sensitivity. The application of this method showed good fit to natural humanreaching movement characteristics suggesting, not very unexpectedly, that humans did avoid uncomfortable joint configurations.

There are clearly a few problems with this method. One of them is the mentioned low sensitivity of the discomfort scores to joint rotation in the midrange. The second is that discomfort ratings for one of the joints may depend on positions (motion) in other joints, for instance due to the action of biarticular muscles. There also seems to be a problem with circular logic: During the lifetime, humans make numerous movements while naturally trying to avoid discomfort. As a consequence, joint positions perceived as comfortable are close to the ones used most frequently during typical movements.

Many cost functions produce movements that look realistic, suggesting that natural human movements do not violate any of those criteria by much. Whether one of the criteria dominates for specific motor tasks remains an open question. Uniting several criteria into a single cost function with modifiable weights makes the model even more powerful and able to account for movement patterns over a broader range of tasks (Seif-Naraghi and Winters, 1990).

A particular complex cost function addressing both state redundancy and trajectory redundancy at the level of movement kinematics has been developed based on ideas of posture planning (Rosenbaum et al., 1993). Within this model, first, a search within the space of joint postures stored in memory is performed, and the total cost is computed for each configuration as the sum of the travel cost to this configuration and resultant spatial error cost. The travel cost is defined as $C = \sum_{i=1}^{n} \xi_i(\alpha_i/MT)$, where n is the number of joints involved in the planned movement, α is the angular displacement in a joint, MT is movement time, and ξ is a factor that defines cost of moving a joint. This factor reflects many parameters including inertial load, apparent stiffness, damping, etc. Further a single target posture is selected as a weighted average of all the postures with preference for postures with smaller total costs (these postures have larger weights). Note that averaging postures in not an obvious procedure, which may result, in particular, in a posture that is anatomically impossible. During the second step, movement into the selected target is performed based on a criterion that defines time profiles of instantaneous velocities in individual joints as functions of their distance to the target posture and also an inertia-related index.

To summarize, criteria based on the ideas of optimality are able to account for certain features of natural movements. These criteria, however, are somewhat arbitrary. They commonly involve computations that require knowing movement time before the movement initiation, which is a questionable assumption.

3.1.2 Inverse optimization

The reviewed optimization approaches are sometimes addressed as *direct optimization*. They all share a common feature: Accepting a cost function based on intuitive considerations, frequently biased by personal views and preferences of individual researchers. Indeed, minimizing energy expenditure, fatigue, discomfort, joint wear, etc., all sound like reasonable assumptions with respect to natural movements. However, the element of subjectivity has been a weak spot of the approach. Recently, a step has been made toward an objective estimation of a cost function (assuming that it exists). This method, termed *analytical inverse optimization* (ANIO), attempts to find an analytical function that is likely minimized by the system by observing its behavior over a range of tasks.

One of the inverse optimization approaches is based on a two-stage search, first among classes of analytical functions and then among parameters within each class (Bottasso et al., 2006). On the one hand, this method requires a lot of computational power. On the other hand, it has the advantage of searching over large classes of analytical functions. The other method is limited to a particular class of cost functions, namely functions additive with respect to the elemental variables, i.e., those with costs associated with action by an elemental variable being independent of values of other elemental variables (Terekhov et al., 2010). This approach is based on the Lagrange principle, which allows estimating the derivatives of unknown cost functions based on observations of performance over a large range of values of task constraints assuming that each action obeys the same unknown optimization principle.

So far, this method has been applied successfully to the problems of force and moment of force production by parallel kinematic chains in isometric conditions, specifically to multifinger pressing and prehensile tasks (problems of state redundancy). The results of ANIO have shown sensitivity to fatigue, healthy aging, and neurological disorders (Park et al., 2010, 2011, 2012). The have also allowed distinguishing among competing sets of variables as candidates to elemental variables manipulated by the central nervous system (Niu et al., 2012). The method failed, however, when it was applied to force production tasks performed by a serial kinematic chain, e.g., the hand force and moment production on a fixed handle (Xu et al. 2012). Moreover, in the latter case, it has been proven that no additive cost function can successfully account for the experimental observations. Overall, so far, the range of application of ANIO has been relatively narrow.

3.1.3 Optimal feedback control

The goal of *optimal control* is to find a way to control a system, which changes in time, in such a way that certain criteria of optimality are satisfied. Consider a system of interest involving a set of state variables that change in time, **X**(t). The rate of change in these variables, d**X**(t)/dt, is defined by their initial values ($\mathbf{X}_0$ at some instant of time, t = 0) and by the current values of the control variables, **C**(*t*): $\frac{dX(t)}{dt} = f[X(t), C(t), t]$, $\mathbf{X}(0) = \mathbf{X}_0$. **X**(*t*) can be found if one knows **C**(*t*) over a certain time interval, for example, from 0 to T, the initial state of the system $\mathbf{X}_0$ and the form of the function *f*. Now, let us try to define a time profile **C**(*t*) such that a cost function ***J*** is minimized over the same time interval: $J = \int_0^T F[X(t), X(t), t]dt + S[X(T), T]$.

In this expression, *F* is a function reflecting the costs associated with changes in both **C**(*t*) and **X**(*t*). For example, *F* may include any of the cost functions reviewed in earlier sections, e.g., those related to energy expenditure, fatigue, effort, etc. The *S*-function gives the so-called *salvage value* corresponding to the final state **X**(*T*) and makes sure that the control is not only minimizing ***J***, but it also leads to a movement goal. For example, the *S*-function does not allow the system not to move at all. Further constraints can be imposed on each component of ***J***, for example, to limit the **X**(*t*) and **C**(*t*) variables to a range compatible with the human anatomy and physiology.

A method of *optimal feedback control* was developed to address the control of a redundant system of effectors (Todorov and Jordan, 2002). A cost function, similar to the function ***J***, was formulated in such a way that it included a measure of neural effort related to changes in **C** variables and a measure of accuracy of performance. In other words, the controller was expected to apply neural effort (changes in **C**) only if these changes improved accuracy of performance. For quantitative purposes, the two components, those related to neural effort and to accuracy, were converted to commensurate units with appropriately selected coefficients. The former was represented with the variance of the control signals to individual effectors during task execution. The latter was the squared difference between a task-specific performance variable and its required value. Within this model, the controller recomputes a new desired trajectory in the redundant space of elemental variables at every moment in time. Deviations in the trajectories of elemental variables from the previously computed ones that have no effects on performance (i.e., within the self-motion space for the corresponding performance variable) are not corrected, whereas deviations that interfere with important performance characteristics (those within the range space) are.

There are several major advantages offered by the optimal feedback control approach. In particular, this approach offers a method of producing flexible behaviors leading to a desired goal, which is one of the characteristic features of voluntary movements. Indeed, if external forces or variations in the intrinsic state of the body lead to deviations of elemental variables, effort will be spent on correcting only those deviations that affect importance performance characteristics. Other deviations will allow developing and lead to variable involvement of elements, for example, as in the classical Bernstein's study of movements by professional blacksmiths mentioned earlier (Bernstein, 1930). This feature gives the optimal feedback control approach clear advantages as compared with more traditional optimization approaches, which define unique solutions for tasks characterized by motor redundancy. This approach can also make quantitative predictions concerning the distribution of work across multiple effectors and in variation in feedback control with changes in task demands and the correlation structure between different effectors (reviewed in Diedrichsen et al., 2010).

On the other hand, this approach shares a few features with the earlier optimization approaches that may be viewed as drawbacks. Among them is the assumption of neural computations and the necessity to estimate movement time before action is initiated, as well as at all intermediate times during the action, as the computation of the projected cost involves integration until movement termination.

3.2 Systems of coordinate: how to count degrees of freedom

The first issue related to the problem of motor redundancy is defining elemental variables, i.e., those that are produced by elements at the selected level of analysis. This is not a trivial step. For example, some apparent elemental variables may be linked to each other by laws of nature, e.g., those reflecting anatomical or physiological constraints, that are unrelated to the specific task. For example, finger joints typically change proportionally during grasping tasks (Santello et al., 2002; Jerde et al., 2003). Counting motion in each of the finger joints as a kinematic elemental variable may be, therefore, unwarranted. In addition, if an apparent elemental variable has no effect on performance whatsoever, counting it could also inflate the number of effective elemental variables. Of course, overlooking important elemental variables may lead to a mistake in the opposite direction, underestimating their number.

Typical studies of action and perception involve measuring variables that are related to both neural control processes and external conditions of the action. Some of those variables are salient to the task, for example, fingertip trajectory during pointing movements. Other measured variables reflect the outputs of elements that contribute to the performance, such as joint rotations during pointing, digit forces and moments during object manipulation, and muscle activations in multimuscle tasks that may involve a limb or the whole body. Some of the elemental variables are not directly measured but computed based on measured variables and assumed or measured parameters of the system; these include, for example, joint torques and muscle forces.

There have been several attempts to make a step toward identifying elemental variables as axes in a system of coordinates reflecting neural processes based on measured joint rotations, forces, and muscle activation. In his classical studies, Bernstein (1947, 1967) suggested that the central nervous system united elements into groups with correlated changes in elemental variables. The purpose of such an organization might be reducing the number of parameters manipulated at task-related neural levels. This idea has been developed more recently using methods of correlation analysis within large sets of measured variables produced by apparent elements such as joints, digits, and muscles. Several such methods have been used to study "*muscle synergies*" defined as relatively stable groups of muscles with correlated changes in activation levels (D'Avella et al. 2003; Ivanenko et al., 2004; Tresch and Jarc, 2009). In other studies, a different term has been used, *muscle modes* (Krishnamoorthy et al., 2003a). There are more related terms, which are defined relatively loosely, such as *modules* and *primitives*. These terms imply building blocks for natural actions but typically do not offer broadly applicable operational definitions, i.e., methods of identifying such blocks.

The word *synergy* is one of the most frequently used and least frequently defined in the field of movement studies (for recent reviews see Latash and Zatsiorsky, 2016; Bruton and O'Dwyer, 2018). Using the same word in different meanings leads to much confusion. For example, in the clinical literature, this word has meant pathological, stereotypical patterns of motion or muscle activation interfering with intentional movements. Such patterns are frequently seen after stroke and some other disorders (DeWald et al., 1995).

The original meaning of the Greek word *synergy* is "work together" It is not clear whether individual elemental variables (e.g., groups of muscles) can be assigned any kind of meaningful "work" The same is true about pathological synergies, which seem to have no functional purpose; rather, they interfere with meaningful functions. To be consistent with the Greek meaning of this word, we will use it only with respect to coordinated actions of multiple elemental variables that "work together" to stabilize salient performance variables. This topic will be discussed in more detail later.

Synergies or *modes* are typically assumed to be used across tasks and conditions with only relatively minor adjustments. A different system of coordinates, the so-called *control coordinates*, had been introduced recently assuming that these coordinates are task-specific and selected to allow using local control along each of the coordinates (Akulin et al., 2019). This means that a change along one of control coordinates can be corrected by an action directed along the same coordinate only. In this sense, they are similar to the notion of eigenmovements introduced for whole-body actions based on mechanical analysis of multijoint actions (Alexandrov et al., 2001) and proven to be able to ensure stable control in a humanoid robot (Alexandrov et al., 2017). The number of control coordinates can be larger or smaller compared with the number of elements or modes.

This brief review suggests that counting degrees of freedom at the level of control is very much nontrivial. To accomplish this task, one has to accept a particular view on how the neural control of action is organized. This is going to be the main topic of the next part of the book. Meanwhile, let us consider an alternative approach to the apparent problem of dealing with numerous elements at different levels of analysis.

3.3 Principle of abundance

The alternative view at the apparently redundant design of the human body is relatively young (reviewed in Latash 2012; 2016). It starts from a philosophically different attitude described earlier in this section: If a feature of the design of a biological system looks suboptimal, one has to think more and try to find out what evolutionary advantage is afforded by this feature. The apparent redundancy of the human body is not the only feature that looks suboptimal from the point of view of contemporary engineering. A few examples of apparent disadvantages of the human body compared with features of machine design have already been mentioned such as sluggish muscles, fuzzy receptors, and long transmission delays.

If one accepts the creationist approach, it would seem that our bodies were designed by a very poorly trained engineer who placed an ultimate computing machine—the brain—into our skulls to handle all the seemingly unavoidable problems associated with this faulty design. In particular, during quick actions and unexpected perturbations, the brain is expected to perform computations of requisite changes in muscle activations and forces within a few milliseconds. This is a very strong assumption! Indeed, consider a very simple task: Flex the right index finger as quickly as possible following a signal, for example, a flash of red light. After some training, people can initiate this task with a delay of about 100−150 ms. Now let us make the task just a bit more complex and ask the subject to flex the finger to the red light and extend it to the green light. The time delay will increase by about 100 ms! This is how much extra time it takes for the brain to solve the most trivial task: Moving the finger up or down in otherwise perfectly predictable conditions. So, let us not exaggerate the speed at which the brain can "compute." The last word is in quotation marks to emphasize the author's view that the brain, as a product of the evolution, does not compute anything—ever! It can transform variables into other variables, but calling this "computation" is equivalent to saying that the pot with water placed on the stove computes how to transform the heat from the burner to motion of water molecules and then to physical and chemical transformations in a potato placed in the pot. The fact that humans can learn mathematics and design computers is a product of civilization, which should not be confused with evolution. For example, animals frequently outperform humans in motor tasks. However, even the best circus performers among animals have not been able to perform the simplest computational operations.

If one combines the evolutionary approach with physics, the basic assumptions become very different and the apparently suboptimal features of the body require not an omnipotent computer in the skull but deeper analysis of possible advantages of these specific features. An attempt to consider the apparently redundant design of the human body as a useful feature has been made in the form of the *principle of abundance* (Latash 2012). The words redundancy and abundance both mean "something extra" but they differ in viewing it as a source of problems (or something to be used only in emergency) versus something one enjoys, e.g., a useful design with a particular purpose.

According to the principle of abundance, the apparently redundant elemental variables are not eliminated during natural actions but used to stabilize performance variables salient for the task. The word *stabilize* is used here in its direct meaning as an ability to return quickly to a desired state or trajectory under such factors as different initial conditions, spontaneously varying intrinsic body states, and unexpected changes in external forces.

Imagine that you are walking along a hallway with a cup of coffee in the hand. Of course, you would like the cup to be vertical or very close to vertical at all times. Note that the cup orientation is affected by motion of all the major arm joints as well as of the legs and trunk. This is an apparently redundant set of kinematic elemental variables that all contribute to a single important variable, orientation of the cup with respect to the vertical. If you "eliminated redundant degrees of freedom" and selected a single optimal solution for this task, any spontaneous deviation of any of the joints from the optimal solution would lead to a change in the cup orientation, potentially resulting in spilled coffee. The same is expected if someone suddenly pushes the elbow of your arm with the cup. The same will also happen if you decide to do something else with the same effector: perform a secondary task, which may not be compatible with the selected, apparently optimal, solution.

Let us also keep in mind that every step is associated with not perfectly predictable reaction forces from the ground on the feet that propagate through the multilink body and act on all its segments. These forces produce perturbations of the arm with the cup of coffee at every step. This simple example suggests that walking with a cup of coffee along the hallway requires ensuring dynamic stability of the cup orientation. Using peripheral receptors to identify errors in cup orientation and then issuing corrections for the ongoing action is not feasible given the aforementioned long conduction time delays. So, stability of cup orientation during this action has to be organized within the central nervous system.

The principle of abundance suggests that humans use the central nervous system to unite all the elemental variables (e.g., joint rotations) into an organization that allows the elemental variables to move relatively freely in directions that do not affect the salient performance variable (such as cup orientation) while resisting motion in directions that change the cup orientation. In other words, the central nervous system ensures high stability of the elemental variables in directions that tilt the cup while allowing low stability in other directions. Such an organization is expected to ensure stable performance of the task under various complicating effects such as those associated with deviations of joints from their typical trajectories due to either external perturbations or spontaneous variations of body states.

Imagine now that you walk down the same hallway with another object in the hand, for example, a tennis ball, which does not require keeping its orientation constant at all times. If someone pushes your elbow, the arm will move differently compared with the cup-with-coffee example: This motion will not preserve the tennis ball orientation with respect to the vertical. This means that stability of performance variables is not something inherent to the central nervous system but is a task-specific feature of behavior: We can modify stability of performance variables as we wish (Schöner, 1995).

Stability in abundant systems

4

Stability of biological movement is arguably its most important feature. Unstable movements are useless given the mentioned features of the animal body such as relatively slow muscles and long transmission delays of relevant neural signals. Imagine, for example, that you step unexpectedly on a stone during walking. The forces between the foot and the ground will change abruptly as compared with those expected from stepping on a smooth, flat surface. These forces produce perturbations for individual leg and trunk joints, as well as for the rest of the body. Dealing with such perturbations, in particular during quick walking, is not an easy problem as demonstrated by contemporary bipedal robots, which struggle navigating unpredictable environment. Note that the robots have clear advantages as compared with humans in the power of their torque motors and close to zero transmission time delays. The striking ability of animals, including humans, to navigate unpredictable environments suggests that the neural control of actions is associated with ensuring their *dynamical stability*.

4.1 Definitions and examples of stability

The most direct definition of stability is based on effects of small, brief perturbations on the object of interest. For example, if an object resides in an equilibrium state, the state is called stable if the object comes back to the initial state following a brief, small perturbation. Fig. 4.1A illustrates a ball on the bottom of a pit. If the ball is displaced from the bottom by an external force and then released, it will move back and, assuming some friction between the ball and the surface, come to rest on the bottom. This allows addressing the initial state of the ball as stable equilibrium. If the ball is on the top of a hill (Fig. 4.1B), it may also be in an equilibrium state, but this equilibrium is unstable. Indeed, a small deviation of the ball from the original position will result in its motion away from the initial position.

A similar set of notions can be applied to a moving object. For example, if the same ball rolls down a half-pipe (Fig. 4.1C), small deviations of the trajectory induced by an external brief force application would result in the ball reaching about the same location at the bottom of the half-pipe (assuming some friction). The same is expected if the ball is released from different locations on the top of the half-pipe (i.e., its initial conditions vary across runs). If the half-pipe is turned upside down (Fig. 4.1D), releasing the ball from different locations on its top would lead to different trajectories that never converge. The same is expected from small force pulses applied at any time along the trajectory. The former example illustrates a dynamically stable behavior, while the latter one a dynamically unstable behavior.

Physics of Biological Action and Perception. https://doi.org/10.1016/B978-0-12-819284-9.00004-8

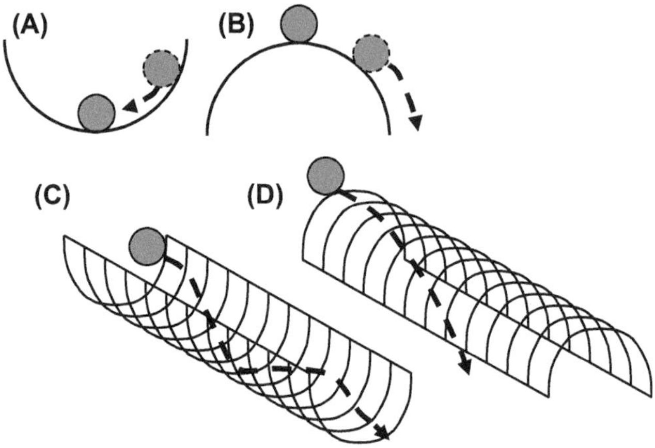

Figure 4.1 Illustrations of stable equilibrium ((A): a ball on the bottom of a pit), unstable equilibrium ((B): a ball on the top of the hill), dynamically stable trajectory ((C): a ball rolling down a half-pipe), and dynamically unstable trajectory ((D): a ball released on the top of a half-pipe turned upside down).

Fig. 4.1 illustrates the behavior of a single object along one or two coordinates. As described in the previous sections, typical animal movements involve multiple elements. Behavior of such systems can be analyzed in a high-dimensional space of elemental variables (those produced by elements) contributing to the behavior and in a lower-dimensional space of a salient performance variable (which can have a few dimensions). Keeping the performance variable at a desired value is equivalent to limiting deviations in the space of elemental variables to directions that lead to no change in that performance variable. The problem of stabilizing a salient performance variable can be formulated in the form of two questions: (1) How can the central nervous system organize and stabilize a subspace corresponding to a desired value of the salient performance variable in the high-dimensional space of elemental variables? (2) How can the unavoidable variance in the elemental variables be channeled primarily into the subspace that does not affect the salient performance variable?

Consider first behavior of an arbitrary system in the multidimensional space of relevant elemental variables. Different directions in that space may have different stability properties. If you ask a person to perform the same action a few times, the initial intrinsic state of that person will vary across repetitive trials. In addition, external forces in the course of the action will also vary. As a result, trajectories across repetitive attempts are expected to differ, converging in dynamically stable directions and diverging in dynamically unstable directions (Fig. 4.2, cf. Fig. 4.1C and D). More stable directions will be characterized by stronger convergence while more unstable directions will show larger divergence of the trajectories. If one identifies a phase along the trajectory and compares intertrial variance of coordinates at that phase, larger variance is expected in less stable directions and smaller variance in more stable directions. This illustration suggests that *intertrial variance* may be used as an index of stability. This insight has been developed into a quantitative tool to analyze stability of natural human actions described in the next section.

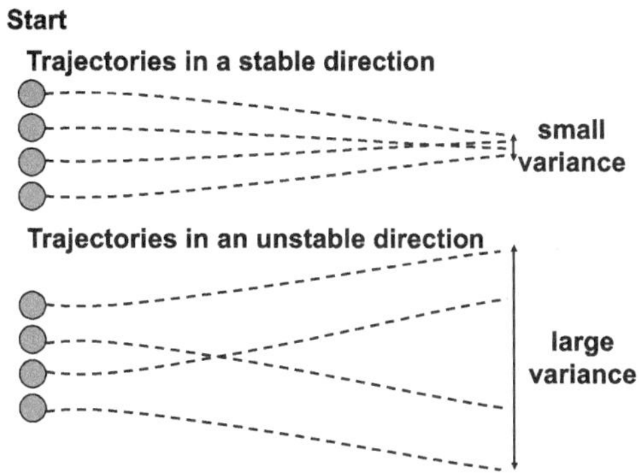

Figure 4.2 Trajectories across repetitive attempts always start in slightly different initial states and are expected to converge in dynamically stable directions and diverge in dynamically unstable directions. This will be reflected in intertrial variance computed at a selected phase along the trajectory.

Imagine now that you place a small spring, for example, a typical spring from a pen, between the thumb and the index finger. Now try to compress the spring quickly along its main axis. (You can actually run this experiment on yourself.) Most likely, the spring will buckle and jump away in an unpredictable direction. You honestly tried to compress the spring along its long axis, and it moved orthogonal to that axis. What could be the reason for this misbehavior? The reason is simple: The spring is relatively stiff (and more stable) along its main axis and relatively compliant (and relatively less stable) orthogonal to that axis. This experiment illustrates another feature of motion in directions of high and low stability. A system is more likely to move along directions of low stability than along directions of high stability. This is true for motions produced by an external to the organism factor, a perturbation, and for motions induced by a quick descending command from higher levels of the central nervous system. This example illustrates another method to quantify stability of a multidimensional system in different directions by quantifying its motion in response to a quick perturbation or a quick action. This method, addressed as analysis of *motor equivalence*, will also be discussed in the next section.

4.2 Task-specific stability and the concept of uncontrolled manifold

A crucial step in the exploration of stability of biological movements was made by Gregor Schöner (1995) about 25 years ago. Schöner suggested that biological systems could stabilize different performance variables in a task-specific manner within the same multidimensional space of elemental variables. In other words, different directions in the space of elemental variables could be made relatively more or less stable by the central nervous system depending on what the actor wanted to accomplish. So, stability becomes not an inherent property of biological action but a controlled one. This idea proved to be very productive leading to the development of tools to quantify stability and application of those tools to movements across conditions, populations, and species.

To make this idea practically useful, one had to develop a method of mapping between high-dimensional spaces of elemental variables and low-dimensional spaces of potentially important performance variables. Such a method was developed based on the concept of uncontrolled manifold (UCM; Scholz and Schöner, 1999). UCM, by definition, is a subspace within the space of elemental variables where a potentially important performance variable does not change. So, if keeping this performance variable at a desired value is the only goal of the central nervous system, it does not have to exert control over the system as long as it resides within the UCM (this is the justification of the term *uncontrolled*). As a result, directions within the UCM are expected to show relatively low stability, whereas directions orthogonal to the UCM (referred further as ORT) are expected to show high stability. Note that these are only expectations based on a researcher's guess on what variable(s) the central nervous system of the subject tries to stabilize in a given task. These expectations do not have to be always confirmed. Indeed, it is possible that the central nervous system cares more about a variable that differs from the one specified explicitly in the task formulation. It is also possible that stability of a performance variable depends on what the actor plans to do; for example, if this variable it to be changed quickly, stabilizing it may be counterproductive.

Consider a very simple task: You are holding a small object, e.g., a shot glass, in the air between the thumb and index finger (Fig. 4.3). To keep the object vertical and motionless, the horizontal forces produced by the two digits should be equal to each other in magnitude so that the resultant horizontal force is zero. This is an abundant system: Two fingers are acting together to produce a certain value of the resultant horizontal force. The space of elemental variables is two-dimensional. The performance variable is one-dimensional leaving one more dimension for the UCM. Two slanted lines in Fig. 4.3 show the UCM where the resultant force stays zero

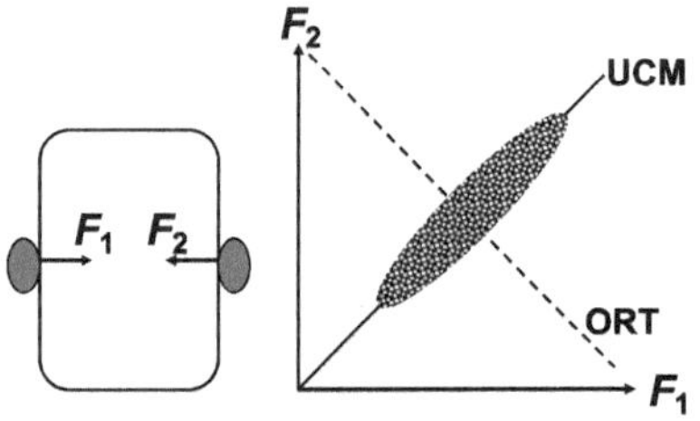

Figure 4.3 Holding the object with two digits (left drawing): The horizontal forces produced by the digits should be equal to each other in magnitude so that the resultant horizontal force is zero. In the right graph, the two *slanted lines* show the uncontrolled manifold (UCM) where the resultant force stays zero (*solid line*) and orthogonal to the UCM direction where the resultant force deviates from zero (*dashed line*, ORT). Note that absolute force magnitudes are plotted. Higher intertrial variance may be expected along the UCM compared to variance along ORT (shown by a cloud of data points).

(solid line) and orthogonal to the UCM direction where the resultant force deviates from zero (dashed line, ORT). Let us now imagine that the subject is asked to perform two tasks. The first is picking up the object and holding it in the air multiple times. Individual digit forces are measured in each trial and a corresponding point is placed on the force–force space. Of course, nobody is perfect, so the subject is expected to show some intertrial variance in the resultant force, for example, due to small deviations of the object from the vertical. This variance will translate into deviations of the recorded points along the dashed line (ORT direction) in Fig. 4.3. What deviations are expected along the UCM? Note that, by definition, these deviations have no effect on the salient performance variable, resultant force. So, the subject can be equally accurate by showing deviations along the UCM that are smaller than, equal to, or larger than those along the ORT. If the subject indeed cares about producing zero resultant force more than anything else, one can expect larger stability along the ORT direction and lower stability along the UCM. This is expected to lead to higher intertrial variance along the UCM compared with variance along ORT (see the ellipse of data points in Fig. 4.3).

In the example illustrated in Fig. 4.3, comparison of variances along the UCM and ORT directions is easy because each of them is one-dimensional. More typically, the dimensionalities of the UCM and ORT subspaces are different. This does not present a major problem, however, because variance is an additive measure of variability. This means that, if you have a system of several elements that produce elemental variables, which sum up to produce a performance variable, adding a new element with some variance, V_{NEW}, is expected to lead to an increase in the variance of the performance variable by V_{NEW} assuming that the new element does not covary its output with the outputs of other elements. So, comparison of variance indices across subspaces of different dimensionalities can be done using indices of variance quantified per dimension.

There is a more serious potential problem. Variance is a linear measure of variability. In the previous example, the two digit forces summed up to produce the resultant force. This corresponds to a linear equation that links the two elemental variables with the performance variable. While such cases happen, more frequently, systems of elemental variables produce performance variables via nonlinear transformations. For example, during fingertip movement in space, fingertip trajectory is linked to individual joint rotations (which may be viewed as elemental variables) via functions involving sines and cosines of joint angles (see Zatsiorsky, 1998). As a result, the UCM in the joint angle space corresponding to a fixed coordinate of the fingertip is curved. To allow analysis of intertrial variance, such a system has to be linearized, which is not a trivial operation that may be legal only assuming small intertrial variability in the joint configuration space. Fig. 4.4 illustrates pointing motion by a four-link "arm" with three joints. If this two-dimensional "subject" performs the same pointing task multiple times, for any phase over the movement duration, the UCM can be computed corresponding to the average across-trial joint configuration at that particular phase. Such UCMs for different phases are illustrated in Fig. 4.4 with dashed lines. They are curved. However, within a small vicinity of the average joint configuration, the curvature is small, and the UCMs can be reasonably

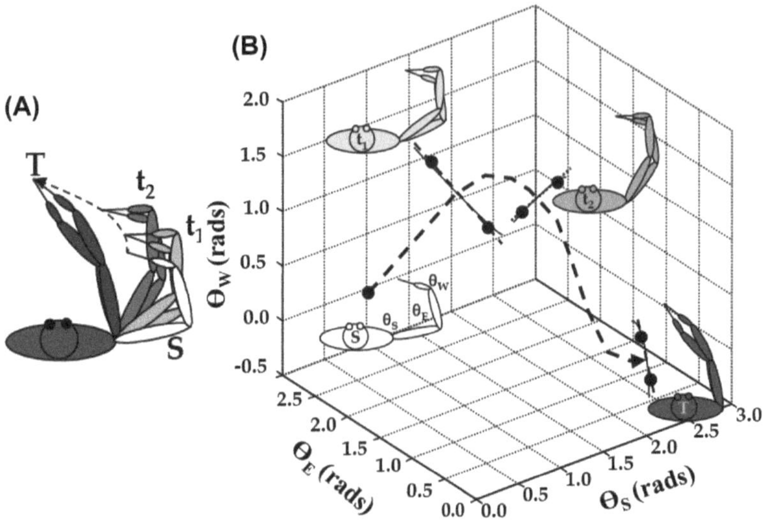

Figure 4.4 (A): Pointing motion performed in a plane by a four-link "arm" with three joints. (B): Average trajectory across trials is shown as a thick *dashed line*. At selected phases, joint configurations are shown for two individual trials. The uncontrolled manifolds (UCMs) for those phases are shown with *dashed lines*, while their linear approximations are shown with *solid lines*. They nearly overlap.
Reproduced by permission from Latash, ML., Scholz, JP., Schöner, G., 2007. Toward a new theory of motor synergies. Motor Control 11/3:276–308. https://doi.org/10.1123/mcj.11.3.276. (© Human Kinetics, Inc.).

approximated with straight lines (solid thin lines in Fig. 4.4). This operation substitutes the UCM with its linear representation.

Such operations are typically done using the notion of the *Jacobian matrix* (**J**). This is a matrix consisting of partial derivatives of the performance variable (*PV*) with respect to each of the elemental variables (*EV*): $\mathbf{J} = |\partial \mathbf{PV}/\partial \mathbf{EV}|$.

Then, the UCM can be approximated with the *null-space* of **J**, which is defined as the subspace where small deviations of elemental variables lead to no change in the performance variable. This definition of the null-space makes it very similar to the UCM. There is only one important difference. The UCM reflects the actual system of interest with the actual, possibly nonlinear, relations between changes in elemental variables and resulting changes in the performance variable, while the null-space of **J** reflects a simplified system where such relations have been linearized (reviewed in Latash et al., 2007).

Now let us consider a different task performed by a person holding an object between the thumb and index finger (cf. Fig. 4.3). Imagine that the subject is asked to move the handheld object quickly to a new location along the line connecting the point of digit contact and stop at the new location (Fig. 4.5). During the motion, the resultant force will have to change to produce the object acceleration and deceleration, but after the movement, it has to return close to zero. Such a trajectory is shown schematically in the right graph in Fig. 4.5 in the space of digit forces. Of course, the resultant force may end up with a value that deviates from the initial point in the ORT

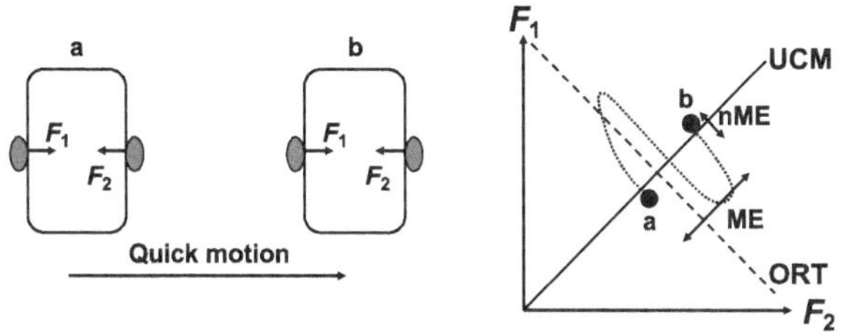

Figure 4.5 The subject moves the handheld object quickly to a new location, from point *a* to point *b*. During the motion, the resultant force will show large changes generating the object acceleration and deceleration. After the movement, the resultant force has to return close to zero. Such a trajectory is shown schematically with the *dotted line* in the right graph in the space of digit forces. At the final coordinate, there should be small deviations along the ORT (nonmotor equivalent, nME), while deviations along the UCM (motor equivalent, ME) can be of any magnitude. Typically, ME > nME.

direction, but not by much. What would be the magnitude of deviation of the final state from the initial state along the UCM? If the system is less stable along the UCM, a larger deviation is expected, similar to what happens in the earlier example of the spring compressed by the index finger and the thumb.

Deviation along the UCM has been termed *motor equivalent*, while deviation along the ORT has been addressed as *non-motor equivalent* (Scholz et al., 2007; Mattos et al., 2011). Motor equivalent deviations are, by definition, wasteful because they lead to no change in the salient performance variable. However, as multiple experiments have shown, these deviations are typically very large (Mattos et al., 2011, 2013, 2015). Non-motor equivalent deviations reflect the task, which may or may not require a net change in the salient performance variable. Of course, if a large change in the performance variable is required, non-motor equivalent deviations are expected to be comparably large. But motor equivalent deviations are not related to the task, only to relative stability of the system along the UCM, which is expected to be relatively low for salient performance variables.

There seems to be a clear advantage of using analysis of motor equivalence to quantify stability of actions by multielement systems as compared with analysis of intertrial variance. Indeed, quantifying variance requires analysis of clouds of data points and, therefore, multiple trials performed by the subject. In contrast, motor equivalence can be quantified in a single trial. Of course, using a single measurement is not expected to produce a reliable estimate of motor equivalence. Still, this method is promising, in particular in clinical studies that involve patients who may have troubles performing the same task multiple times (Freitas et al., 2019).

4.3 Synergies stabilizing action

It is time to introduce the promised definition for a performance-stabilizing synergy: *Synergy is a neural organization of a set of elemental variables with the purpose to stabilize a task-specific, salient performance variable.* According to this definition, elemental variables within a synergy "work together" to achieve a clear and functionally important goal: ensure stability of performance with respect to salient variable(s).

This definition makes methods of analysis described in the previous section directly relevant for analysis of synergies. Indeed, if a synergy stabilizes a performance variable, across-trial variance within the respective UCM (V_{UCM}) has to be larger than variance orthogonal to the UCM (V_{ORT}), both quantified per dimension in the UCM and ORT spaces. In addition, if a subject performs a cyclical action, cycle-to-cycle deviations quantified between comparable phases within the UCM (motor equivalent) are expected to be larger than within the ORT (non-motor equivalent), both indices should also be properly quantified given the actual dimensionality of the UCM and ORT.

If a task explicitly requires accurate performance with respect to a particular variable produced by a set of elemental variables, it is natural to assume that a synergy would stabilize that performance variable. Indeed, quite a few studies documented such synergies across tasks and levels of analysis (reviewed in Latash, 2008). Other studies, however, showed that the central nervous system of the person performing the action might reinterpret the task and stabilize a different performance variable that is not explicitly required by the task. For example, one of the early series of studies (Latash et al., 2001; Scholz et al., 2002) required the subjects to produce cyclical total force changes between two targets at a comfortable rate (paced by a metronome) by pressing with two, three, or four fingers of a hand on individual force sensors. The task was always abundant given that a single performance variable had to be produced by the set of fingers. The only visual feedback the subjects got was on the current total force. To the surprise of the experimenters, the subjects showed multifinger synergies (operationally defined as $V_{UCM} > V_{ORT}$) stabilizing total force only within a relatively small range of force magnitudes. In contrast, they stabilized the total moment of force with respect to the midpoint of the involved fingers (i.e., the rotational action produced by the fingers in pronation/supination), despite the fact that moment of force was not mentioned as part of the task and no feedback on moment of force was provided.

This result was particularly striking for the task performed with two fingers because, for that task, stabilizing moment of force and total force together was impossible. Indeed, as illustrated in Fig. 4.6, the UCM for total force coincides with the ORT for total moment of force, and vice versa. When faced with this dilemma, the subjects consistently preferred stabilizing total moment of force at the expense of destabilizing the total force, which was the explicit task-related variable. Their data clouds were consistently elongated along the UCM for the total moment as shown in Fig. 4.6. These findings were interpreted by the authors as reflecting the lifetime experience with everyday prehensile actions, which typically require accurate production of moment of force, whereas the total pressing finger force may vary within a relatively wide range. For example, during taking a sip from a glass, the rotational action has to be very accurate, but the gripping force can vary within a broad range, from the slipping threshold to the crushing threshold.

So, application of the UCM-based analysis does not guarantee that a specific structure of intertrial variance (or a specific signature of motor equivalence analysis)

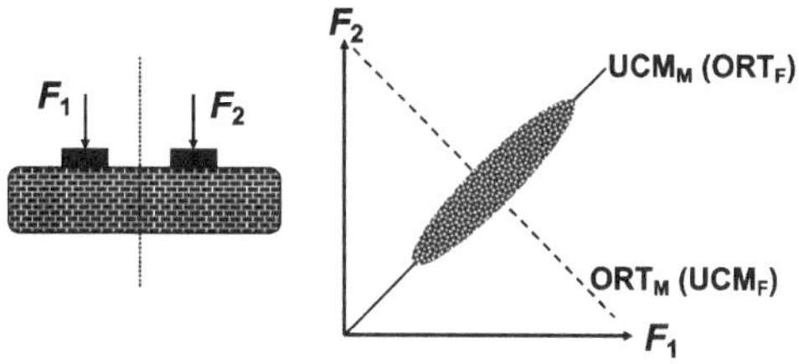

Figure 4.6 The task is to produce a certain magnitude of total force by pressing with two fingers of a hand on two force sensors (left). The uncontrolled manifold (UCM) for total force coincides with the ORT for total moment of force computed with respect to the axis passing in between the two fingers, and vice versa (shown with the *slanted dashed* and *solid lines* in the right graph). So, both total force and total moment cannot be stabilized simultaneously. When faced with this dilemma, the subjects consistently preferred stabilizing total moment of force at the expense of destabilizing the explicit task-related variable, total force, as illustrated with the cloud of data points elongated along UCM_M (ORT_F).

will be discovered. This feature of the analysis opens an exciting opportunity to use it as a probe into strategies used by the central nervous system when it deals with motor tasks. Indeed, these methods allow asking the central nervous system questions such as: Do you care about stability of such-and-such performance variable? Which of the competing performance variables are perceived as most salient given the task? and so on.

4.3.1 Kinetic, kinematic, and muscle activation synergies

Synergies stabilizing action have been explored across a large variety of tasks and systems of elemental variables. Selecting an appropriate set of elemental variables has typically been the most crucial step in the analysis. First, elemental variables within the set should be potentially independent. This means that the central nervous system should be able to change them one by one. Otherwise, clouds of data points in the space of elemental variables are expected to be nonspherical and elongated in certain directions, independently of what the subject tries to do. By pure chance, these directions may be parallel to the UCM or ORT for a particular performance variable leading to false conclusions on a synergy stabilizing that variable being or not being present.

This requirement is rarely met for variables measured in typical experiments. For example, during multifinger tasks, pressing with a finger is associated with unintentional force production by other fingers of the hand. This phenomenon has been addressed as limited finger independence or finger enslaving (Zatsiorsky et al., 2000; Schieber and Santello, 2004). This means that finger forces do not qualify as elemental variables.

During multijoint movements, elemental variables have typically been associated with individual joint angles. This is also not trivial, even during slow actions. Joints are coupled by biarticular and polyarticular muscles, as well as by interjoint reflexes. So, moving a joint is expected to lead to non-zero motion of other joints.

However, introspection suggests that we can move one joint at a time. So far, this intuitive feeling has not been checked in a well-controlled experiment. Sometimes, selection of elemental variables leads to their nonindependence due to purely mechanical factors. For example, angles made by moving segments within an external reference frame (segmental angles) are nonindependent: If a proximal segment moves, its segmental angles change together with the segmental angles of more distal segments to preserve body integrity (Fig. 4.7).

During analysis of muscle activations within a large set of muscles, it would be naïve to expect individual muscle activation levels to qualify as elemental variables (it is time to remind a classical phrase by Hughlings Jackson (1889): "The brain knows nothing about muscles, it knows only movements"). Indeed, muscles are commonly united into groups (*muscle modes* or *muscle synergies*, see earlier) with correlated changes in activation levels within a group.

In many experimental studies of synergies, elemental variables have been associated with *modes*, defined as variables that can be changed by the controller one at a time and are able to account for the observed data sets. For example, finger modes are defined as variables reflecting intended involvement of a finger leading to changes in forces produced by all the fingers of the hand (Danion et al. 2003). Muscle modes are defined as groups of muscles with parallel changes in activation seen across tasks (Krishnamoorthy et al., 2003a). The mapping between the modes and the original set of measured variables (see Fig. 4.8) can be discovered using various, relatively straightforward, computational procedures.

After elemental variables have been defined, one has to discover how their small changes affect the selected performance variable. In other words, one has to define the Jacobian matrix of the system. This may be trivial for performance variables linked to elemental variables by laws of classical mechanics. Examples include total force or total moment of force produced by a set of fingers and trajectory of the endpoint of a limb resulting from individual joint rotations. Sometimes, however, elemental variables and performance variables are expressed in different units. For example, muscle mode magnitude is expressed in units of muscle activation integrated over reasonably selected time windows; so, mode units are millivolt·seconds (mVs).

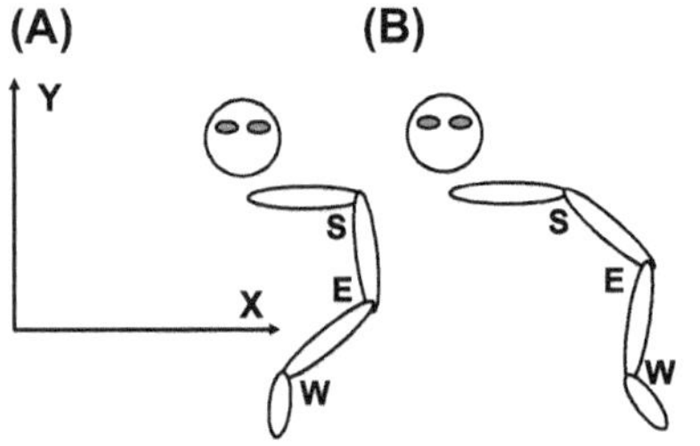

Figure 4.7 The drawings show a two-dimensional "arm" with three joint angles, shoulder (S), elbow (E), and wrist (W). Moving only the shoulder joint (compare the drawings *A* and *B*) leads to changes in all the angles made by each of the three segments within an external reference frame (segmental angles with axes X and Y).

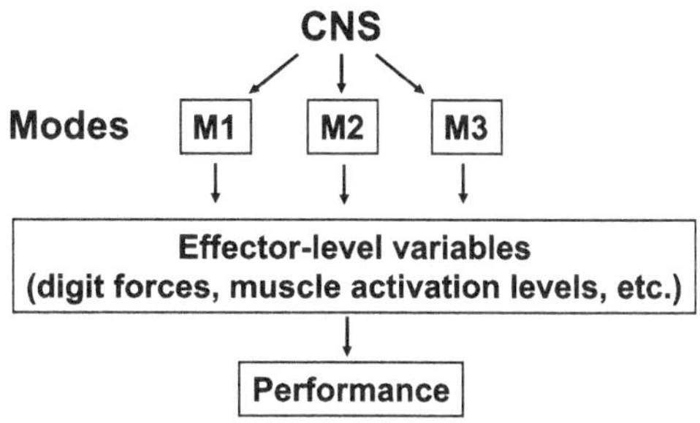

Figure 4.8 The central nervous system is presumed to manipulate a set of modes, hypothetical variables that can be changed one at a time. Changes in modes lead to changes in the output variables of effectors typically recorded in experiments, which lead to changes in salient performance variables. Of course, there are also effects of external forces on performance variables (not shown in the scheme).

How changes in these elemental variables produce changes in action mechanics, measured in newtons and meters, cannot be easily computed and has to be discovered. This is typically done using statistical tools such as multiple regression analysis.

As of now, the number of studies exploring action-stabilizing synergies is on the order of a few hundred (reviewed in Latash, 2008, 2017). These studies have documented synergies that stabilize performance within a range of tasks. These involve constrained laboratory tasks such as swaying to the metronome while standing, producing a combination of force and moment while pressing on force sensors, and pointing at a target with a pointer. Other explored tasks were more natural and could be viewed as building blocks for the repertoire of natural actions, for example, sit-to-stand and stand-to-sit actions, making a step, and moving a handheld object. Still other studies explored highly specialized actions such as quick-draw shooting a pistol at the target (Scholz et al., 2000) and throwing a frisbee into the target (Yang and Scholz, 2005). While most studies involved young, healthy participants, synergies were also quantified in persons with atypical development (Down syndrome), healthy elderly, and patients with a range of neurological disorders (reviewed in Latash, 2008; Latash and Huang, 2015). There were also attempts at quantifying synergies in animals including locomotor tasks in cats and targeted upper extremity actions in monkeys (e.g., Klishko et al., 2014; Mangalam et al., 2018).

Structure of variance at the level of elemental variables is a rather sensitive tool as demonstrated in studies of the typical postural sway during quiet standing. When young, healthy people stand quietly, there are continuous changes in many mechanical variables addressed collectively as postural sway. Frequently, sway is quantified using time changes in the coordinates of the center of pressure (the point of application of the vertical resultant force acting on the body) and center of mass (COM) of the body. One of the influential hypotheses has been that, during quiet stance, the body oscillates about the ankle joints as an inverted pendulum (Winter et al., 1996). If this is true, COM displacements are defined by a single kinematic variable, ankle joint displacement, and there is no possibility of organizing a kinematic synergy in the joint configuration space stabilizing the COM trajectory. A study of individual joint displacements along the body axis has shown, however, that most of the joint configuration variance was within the UCM computed for the COM trajectory (Hsu et al., 2007). A later study has confirmed this finding using a multijoint model and analyzing it in the muscle activation space (Reimann and Schöner, 2017). In that study, the authors found that control laws that actively coordinate muscle activations among the individual joints generated variance patterns compatible with the experimentally observed patterns of COM stabilization, while control laws that control each joint separately did not.

The latter finding may be a consequence of a more general rule that, within a hierarchical system, synergies at different hierarchical levels compete with each other. This rule was suggested based on a series of experimental studies of multidigit synergies during pressing and prehensile tasks (Gorniak et al., 2007, 2009). For example, in one of those studies, strong between-hand synergies were observed stabilizing total force produced by four fingers, the index and middle fingers of each hand, pressing on individual force sensors. There were, however, no two-finger synergies within each of the hands that would stabilize the force produced by that hand. This result is not 100% unexpected. Indeed, at the two-hand level of analysis, large variance within the UCM (V_{UCM}) is a sign of a strong two-hand synergy. The large V_{UCM}, however, requires comparably large variance of each of the individual hand forces. However, large hand force variance means comparably high variance along ORT at the within-a-hand analysis. Therefore, there is an inherent trade-off between synergies at the two hierarchical levels. This trade-off is not absolute: Synergies at both levels can coexist. This requires, however, that the inequality $V_{UCM} > V_{ORT}$ at the upper level (two-hand level in the described example) is associated with relatively low V_{UCM} magnitudes, which would leave room for this inequality at the lower (within-a-hand) level.

4.3.2 Anticipatory control of stability

While action stability is crucial for many everyday behaviors, it may be counterproductive if the task requires changing a performance variable quickly. Indeed, if one assumes that there is a dedicated neuronal structure responsible for stabilization of a performance variable, this structure is expected to correct any deviations of that variable from a desired value. Such deviations can be produced by changes in external forces. They can also be produced by an input from other structures within the central nervous system, for example, those that generate neural control signal for a quick action associated with a change in that performance variable. In either case, the synergy is expected to resist changes in the performance variable and, as a result, slow down the planned action or even prevent it altogether.

Studies of synergies during voluntary actions at a range of speeds have shown that, indeed, indices of synergies stabilizing a variable drop with an increase in the rate of change of that variable (Latash et al. 2002; Friedman et al., 2009). This is primarily reflected in a disproportionate increase in the component of intertrial variance orthogonal to the UCM. During very fast movements, synergies can disappear altogether.

When a person prepares to perform a quick action from steady state, changes in the synergy index for the salient performance variable can be seen as early as 300−500 ms before the action initiation. This phenomenon, known as anticipatory synergy adjustment (Olafsdottir et al., 2005; Shim et al. 2005; Togo and Imamizu, 2016) is illustrated in Fig. 4.9. Assume that two elements are performing a task that requires producing a certain magnitude of their sum, e.g., pressing with two fingers on force sensors to produce a desired magnitude of total force. During steady-state total force production, typically, a strong synergy is seen stabilizing total force. This is reflected in relatively low intertrial variance leading to total force change (along ORT direction) as compared with variance along the UCM. In the force−force plane illustrated in Fig. 4.9, this structure of variance is represented with an ellipse of data points elongated along the UCM.

Now imagine that the person produces a force pulse into a target at a self-selected time. If several trials are performed in a row and the individual trials are aligned by the force pulse initiation time, changes in the two components of the intertrial variance can be seen a few hundred of milliseconds before the pulse initiation. These changes involve an increase in V_{ORT} and/or a drop in V_{UCM} illustrated in the right panel of Fig. 4.9 with the original thin ellipse of data points for the steady-state phase and the thicker ellipse for the time of force pulse initiation, t_0. This is also illustrated as a drop in the relative amount of variance along the UCM (ΔV in the left graph).

The phenomenon of ASA is also illustrated in Fig. 4.10 using a more abstract example. Imagine an object in a stable steady state as residing on the bottom of a deep potential well. Imagine now that the person in control of the object plans to change its coordinate quickly to a new value (shown as point B in Fig. 4.10).

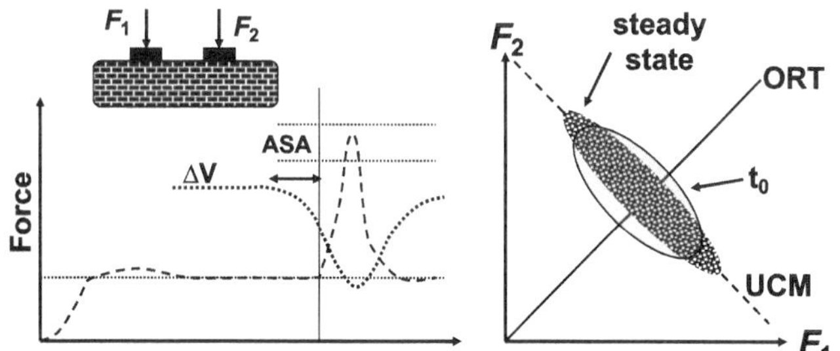

Figure 4.9 A person performs a two-finger accurate force production task multiple times. The task is to follow a target set at a constant total force value and then, at a self-selected time, to produce a quick force pulse into the target window (shown with thin horizontal *dotted lines*). During the steady-state phase, most intertrial variance is within the uncontrolled manifold (UCM) for total force corresponding to a large magnitude of an index (ΔV) reflecting the relative amount of variance within the UCM in the total variance. In preparation to the force pulse, a few hundred ms before its initiation (t_0), ΔV drops. This anticipatory synergy adjustment (ASA) may be associated with an increase in variance along the ORT and/or a decrease in variance along the UCM, illustrated by the two ellipses of data distribution in the right graph.

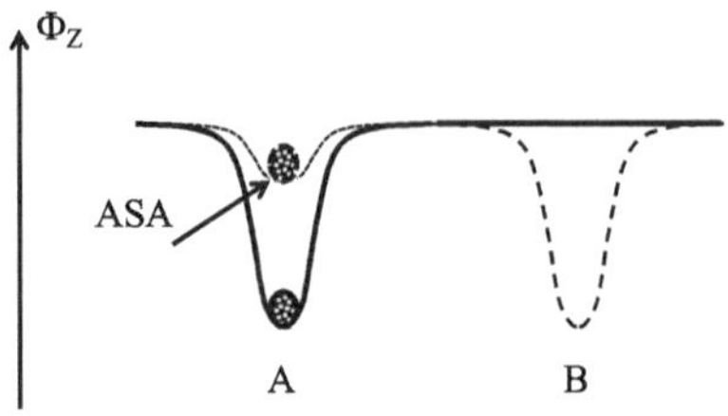

Figure 4.10 An object in a stable steady state resides on the bottom of a deep potential well (a minimum of a potential Φ along the coordinate Z, Φ_Z). If a person plans to change the coordinate of the object quickly (e.g., from point A to point B), during the preparation time, the initial coordinate of the object does not change but the potential well becomes less deep (ASA). This is associated with relative loss of stability while no overt action can be seen.

During the preparation time to the action, the initial coordinate of the object does not change but the potential well becomes less deep. This is associated with relative loss of stability while no overt action can be seen. Of course, when the time of action initiation comes, it is easier to initiate motion of the object to point B from the new, less stable, state.

To appreciate the potential functional importance of ASAs, consider the design of a typical airplane and of a typical bird (I am very much grateful to Zia Hasan for this example!). Both have elongated bodies, wings, and tails. Only airplanes, however, have vertical fins (rudders) on the tail, and some airplanes also have vertical fins at the end of the wings. The purpose of the rudder(s) is to help stabilize the airplane parallel to the ground so that the pilot can concentrate on other aspects of flying. Why hasn't the evolution come up with a similar design? The answer is rather straightforward. For a bird, excessive stability would interfere with agility (maneuverability), which is crucial for survival: A very stable bird is easy prey and unsuccessful predator. ASAs show that the central nervous system in the process of evolution developed the ability to modify stability of salient performance variables in preparation to action. In other words, the evolution came up with the idea of invisible rudders with modifiable size and maybe even shape.

4.3.3 Physiological mechanisms of synergies

Most likely, there is no single mechanism shared by all the various synergies documented for human movements. In two of the following sections, we will discuss the effects of different neurological disorders on synergic control. One of the unexpected lessons from those studies has been that pathological changes in subcortical structures, such as the cerebellum and the basal ganglia, result in significantly impaired synergies (Park et al. 2012; 2013; Jo et al., 2015), while cortical stroke seems to lead to much weaker effects and some studies even claimed no differences in the synergic control following mild cortical stroke (Reisman and Scholz, 2003; Jo et al. 2016). These results suggest that not all the brain structures and loops play equal roles in the control of action stability. Of course, studies in patients, particularly those who suffer from long-standing disorders, should be interpreted with caution. Some of the consequences can reflect adaptive changes in different parts of the central nervous system and/or effects of long-term drug therapy.

A few relatively well-studied mechanisms within the spinal cord may be viewed as synergies stabilizing certain salient variables important for motor actions. Let us consider two examples.

The output cells of the spinal cord, alpha-motoneurons (α-MN), send their output branches, the axons, to muscles. These axons have shorter branches, called collaterals, which make excitatory synapses on small inhibitory interneurons, the *Renshaw cells* (Fig. 4.11A). The Renshaw cells project back onto the α-MN of the pool; there are other projections of these cells, which we will not discuss at this time. As a result, all the α-MN within a pool experience negative feedback effects mediated by all the Renshaw cells. This mechanism is addressed as *recurrent inhibition*. It is important that the Renshaw cells are also under the control by descending neural pathways from the brain, which can modulate the gain of this negative feedback.

Imagine that a pool of α-MNs produces a certain combined level of activity. Now, imagine that, for an unknown reason, one of the α-MNs stops generating action potentials. This introduces a drop in the combined activity level. As soon as this α-MN stops generating action potentials, its Renshaw cells stop receiving excitation and become less active or silent. This drop in the activity of those Renshaw cells leads to a reduction in the amount of inhibition experienced by other α-MNs of the pool causing an increase in their activity levels (Fig. 4.11B). Overall, the effect mediated by the Renshaw cells partly compensates for the original "error", which is a typical feature of negative feedback loops. So, the mechanism of recurrent inhibition contributes to negative covariation in the outputs of the elements of the pool, the α-MNs, and may be seen as mediating a synergy stabilizing total output of the pool.

Let us consider the mechanism of stretch reflex described earlier. The reflex is based on excitatory effects of length-sensitive sensory endings located in a muscle

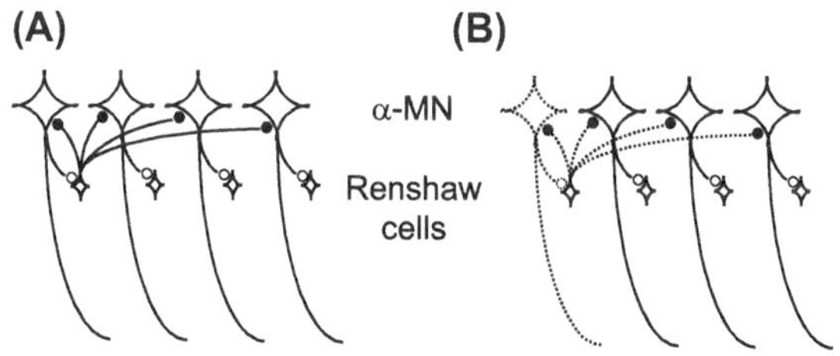

Figure 4.11 The output cells of the spinal cord, alpha-motoneurons (α-MN), send their output branches, the axons, to muscles. These axons have short collaterals, which make excitatory synapses (*open circles*) on small inhibitory interneurons, the Renshaw cells, which project back to all the α-MNs of the pool (shown with *filled circles*) and some other targets. If one of the α-MNs stops generating action potentials, its Renshaw cells stop receiving excitation and become less active or silent (shown in B with *dotted lines*). This leads to a reduction in the amount of inhibition experienced by other α-MN of the pool causing an increase in their activity levels.

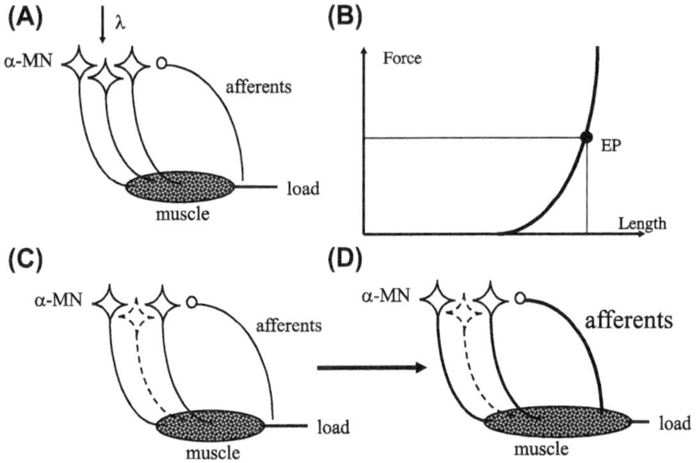

Figure 4.12 The stretch reflex is based on excitatory effects of length-sensitive sensory endings located in a muscle on alpha motoneurons (α-MN) that send signals to the same muscle (A). A muscle controlled by setting a value of the stretch reflex threshold (λ) and acting against an external load comes to an equilibrium at a combination of muscle force and length ((B): the equilibrium point, EP). Imagine that one of the active motoneurons turns off (shown with *dashed lines* in (C)). The level of muscle activation will decrease leading to lower muscle force. As a result, the external load will be larger than muscle force leading to muscle stretch (D). Muscle stretch will lead to higher activity levels of length-sensitive sensory endings, which will produce stronger reflex excitation of the α-MN pool. Overall, this effect will compensate, at least partly, for the original "error" moving the muscle back toward the EP.

(in special structures called *muscle spindles*) on α-MNs, which send signals to the same muscle. Fig. 4.12A illustrates this reflex loop. Imagine that a muscle acts against an external load and comes to an equilibrium at a combination of muscle force and length (the equilibrium point [EP] in Fig. 4.12B). This state is associated with some level of muscle activation caused by signals from a subset of α-MNs. Imagine now that one of those motoneurons, for an unknown reason, turns off (panel C in Fig. 4.12). The level of muscle activation will drop leading to a drop in the muscle force. As a result, the external load will be larger than muscle force leading to muscle stretch and its deviation from the EP (panel D in Fig. 4.12). Muscle stretch will lead to higher activity levels of length-sensitive sensory endings, which will produce stronger reflex excitation of the corresponding α-MN pool. Overall, this effect will compensate, at least partly, for the original "error" moving the muscle back toward the EP. Hence, one can view the stretch reflex loop as a mechanism stabilizing the equilibrium state of the muscle acting against an external load.

These two examples show that some synergic mechanisms may be viewed as generic and common across species. There may be other mechanisms based on negative feedback loops that could be confined to the central nervous system (as in the example of recurrent inhibition) or involve muscles and peripheral sensory endings (as in the example of stretch reflex). A couple of models have been developed showing that such mechanisms are able to lead to typical synergic signatures, for example, the structure of intertrial variance reflecting stability of specific performance variables within the high-dimensional space of elemental variables (Latash et al., 2005; Martin et al., 2009).

The spinal cord is able to stabilize salient variables during more complex actions. This was demonstrated, in particular, in studies of the wiping reflex in spinal frogs. In those preparations, the spinal cord was cut at the upper cervical level and, as a result, the neural control of the body was only possible by spinal structures. The frogs were able to wipe the irritating stimuli off their backs with a coordinated hindlimb motion even when the hindlimb was loaded with a lead bracelet and when one of the major hindlimb joints was blocked (reviewed in Latash, 1993). Recently, the importance of short-latency, spinal responses for stabilization of hand performance has been demonstrated in a study of human subjects (Weiler et al., 2019). These results are readily compatible with earlier studies demonstrating that spinal reflex loops bring about responses in muscles crossing other joints of the limb leading to interjoint coordination effects (reviewed in Nichols, 1994, 2002).

Studies on the role of feedback of different modalities on performance-stabilizing synergies have provided evidence for a very important role played by the visual feedback. For example, if a person is asked to produce a certain level of total force while pressing with the four fingers of a hand on individual sensors, providing visual feedback on the total force magnitude is associated with strong finger synergies stabilizing total force. If the visual feedback is removed and the person is asked to continue performing the task, very quickly drift in the total force is seen, commonly to lower force values (Vaillancourt and Russell, 2002; Ambike et al. 2015; Reschechtko et al., 2017b). This drift, by itself, may be viewed as a sign of lost stability of force magnitude. In addition, experiments have shown that the drift is indeed associated with a drop in the synergy indices such as those reflecting structure of intertrial variance in the finger force space (Parsa et al., 2016, 2017; Ambike et al., 2016b; Reschechtko and Latash, 2017). Note that this happens while the subject continues to receive natural feedback from force-sensitive finger and hand receptors. This feedback, however, is unable to prevent the loss of stability and rather large drift of total force, up to 30% of the initial force level.

On the other hand, studies of more natural actions, such as holding an instrumented handle gripped by all the five digits of the hand, documented strong synergies stabilizing important components of the resultant force and moment vectors (reviewed in Zatsiorky and Latash 2008). In those studies, no feedback on those mechanical variables was provided but the subject could see their hand with the object and tried to maintain its desired orientation in space. These studies show that synergies stabilizing salient kinetic variables can function in the absence of visual feedback on those variables, as long as feedback on other (e.g., kinematic) task-related variables is available.

4.3.4 Impaired control of stability

To be successful across the whole range of all the various actions performed in everyday life, one has to be able to control stability of salient performance variables depending on the nature of the task. A number of conditions, from natural healthy aging to atypical development and to neurological disorders, lead to impairments in the basic abilities to ensure stability of task-specific salient performance variables during steady-state tasks and to attenuate stability in preparation to a quick action. These have been addressed as *impaired stability* and *impaired agility*, respectively (reviewed in Latash and Huang, 2015). We will discuss issues of the impaired control and coordination in more detail in Section 5.7. Here let us consider only a few examples of disordered synergic control, potentially pointing at neurophysiological mechanisms of unimpaired synergies.

Relatively modest but significant impairments in both stability and agility of actions are observed in the healthy elderly. These include smaller indices of synergies stabilizing performance across tasks (Shinohara et al., 2004; Olafsdottir et al., 2007a) as well as shorter and smaller ASAs (Olafsdottir et al., 2007b) that may interfere with the ability to initiate fast actions. Healthy aging is associated with neuronal loss at different levels of the central nervous system. It is feasible, therefore, that neurons involved in the (unknown) circuits of synergic control also suffer losses that make these circuits less efficient. During steady-state tasks, these processes are expected to lead to partial loss of dynamical stability and the necessity to monitor ongoing actions with sensory signals leading, in particular, to the well-documented larger reliance of older adults on online sensory, in particular visual, information (Seidler-Dobrin and Stelmach, 1998; Rand et al., 2013).

Qualitatively similar, but significantly larger in magnitude, changes in stability and agility are seen across a number of disorders affecting subcortical brain circuits; these include Parkinson's disease, multiple sclerosis, and multisystem atrophy (Park et al. 2012; 2013; Jo et al., 2015, 2016, 2017). In all the mentioned patients, both stability and agility indices are decreased as compared with age-matched healthy control subjects. Changes in the control of stability are arguably one of the first signs of subcortical abnormalities: They can be detected at very early stages of Parkinson's disease and even in populations at high risk for basal ganglia disorders, such as professional welders (Lewis et al., 2016). During their job-related activities, welders are exposed to fumes containing metals such as manganese. Inhaling these fumes is known to lead to accumulation of manganese in brain structures, including the basal ganglia, which are the locus of problems leading to Parkinson's disease and a number of other disorders in the general population.

Cortical disorders, in particular cortical stroke, lead to problems with agility while stability seems to be relatively preserved even in patients who demonstrate substantial changes in the overall movement patterns (Reisman and Scholz, 2003; Jo et al. 2016). These observations suggest that stability and agility are not two outcomes produced by exactly the same neural mechanisms. This conclusion has also been supported by the observations of significant effects of deep brain stimulation in patients with Parkinson's disease on indices of agility, but not on indices of stability (Falaki et al., 2018). These observations potentially help to explain why patients with Parkinson's disease treated with deep brain stimulation typically show an improvement in their movement patterns without an improvement in postural stability and in the occurrence of falls (Weaver et al., 2009). Indeed, an increase in agility is expected to mitigate one of the cardinal signs of Parkinson's disease, bradykinesia (also addressed as akinesia), which is a combination of signs such as slowness of movements, problems with movement initiation, and with switching among movement patterns. So, patients with deep brain stimulation are expected to move faster, including faster, more normal-looking, gait. However, if postural stability remains compromised, walking fast might not be such a good idea because faster gait is associated with larger postural perturbations, in particular those occurring during foot contact with the ground.

4.3.5 Stability and optimality

The operational definition of stability and the method of its quantitative assessment illustrated in Figs. 4.3 and 4.6 make stability and optimality of action hardly compatible (Park et al., 2010). Indeed, the idea of optimality implies that there is a single optimal solution for a task performed by an abundant system. If the actor is perfectly optimal, this single solution is expected to be repeated across trials. The idea of task-specific stability of salient performance variables, however, implies relatively low stability within the solution space (the UCM), which means that variable solutions are expected across repetitive trials.

Fig. 4.13 illustrates this idea using the example of a two-element system performing the task of accurate production of the sum of the two elemental variables: $EV_1 + EV_2 = C$. For any C, this task is associated with an infinite number of solutions represented by points along the corresponding line with negative slope. For example, an optimal solution may correspond to 50:50 sharing of the task between the two variables. It is illustrated with black dots on each of the UCMs. Imagine now that the subject in this experiment performs each of the illustrated tasks multiple times. If there is a two-element synergy stabilizing the task-specific performance,

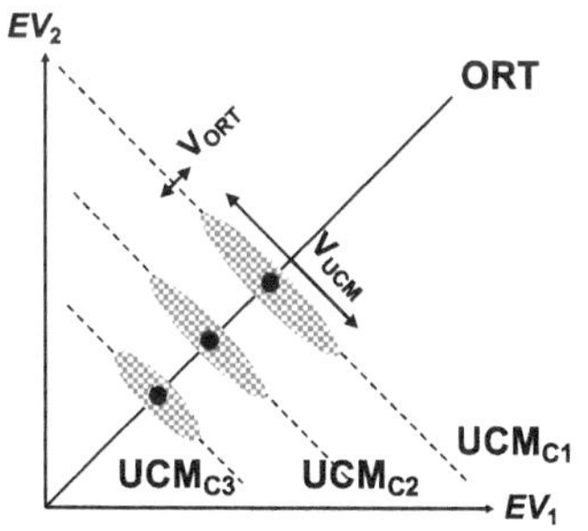

Figure 4.13 An illustration of a two-effector task of producing accurate magnitudes of the sum of their outputs: $(EV_1 + EV_2) = C$. The solution spaces for three magnitudes of C are shown with slanted *dashed lines*. The black dots show optimal solutions (assuming that the 50:50 sharing is optimal). The clouds of data distributions across multiple trials show larger variance along the uncontrolled manifold (UCM) (V_{UCM}) than orthogonal to the UCM (V_{ORT}). Note that large V_{UCM} implies high stability of performance and also leads to proportionally large violations of the optimality principle.

i.e., ($EV_1 + EV_2$), relatively large variability is expected along each of the UCMs compared with variability orthogonal to the UCMs. Note that an increase in deviations of data points from the optimal solution along the UCM implies violations of the optimality criterion and, simultaneously, means higher V_{UCM}, i.e., higher stability of performance. Note also that deviations orthogonal to the UCM (along ORT) do not violate the optimality criterion while hurting stability of performance.

To reconcile the ideas of stability and optimality, it has been suggested that optimality criteria may be used to describe centers for data distributions across trials while the shape of the distributions reflected stability of the action (Park et al., 2010, 2013). Several studies have shown that decline in motor performance with age, fatigue, and neurological disorder (Park et al., 2011, 2012, 2013a,b) is associated with both less optimal and less stable performance. In those studies, the analytical inverse optimization (ANIO, Terekhov et al., 2010, Section 3.1.2) method was applied to data sets during four-finger force/moment production tasks. This allowed identifying subject-specific cost functions. Furthermore, the cost functions were used to predict optimal performance across tasks (a hyperplane in the four-dimensional finger space), which was compared to the actual data sets (another hyperplane in the same space). The dihedral angle between the two planes was used as an index of optimality. This angle was very small in young, healthy adults, on the order of 1−2°, which means that their performance was very close to optimal. It was significantly higher in healthy older adults (about 5°) and in young adults under fatigue (about 7°). It was much higher in persons with neurological subcortical disorders such as Parkinson's disease and multisystem atrophy (15−25°). At the same time, synergy indices computed across repetitive trials at the same task were also the highest in young healthy adults and decreased for the older persons, under fatigue and with neurological disorder. So, while optimality and stability seem to be inherently competitive, both can deteriorate under conditions associated with impaired motor coordination.

Within a young healthy population, however, the inherent competition between optimality and stability seems to reveal itself in preferred patterns of action. Such individual traits in behavior have been traditionally studied in populations with neurological or psychological abnormalities such as autism, dyspraxia, developmental coordination disorder (DCD), and others (Harris et al., 2008; Lingam et al. 2010; Cheng et al., 2017; Curioni et al., 2017). Personality traits such as risk taking, novelty seeking, reward dependence, general arousal, and fear of falling have been linked to cerebellar activity (Picerni et al., 2013; Petrosini et al., 2015) and are known to correlate with indices of performance (Miller and Saygin, 2013; Zaback et al., 2015). Individual traits across young healthy persons have also been reported in feedforward force control tasks as reflected in indices of anticipatory synergy adjustments (ASAs, Togo and Imamizu 2016; Imamizu et al., 1998).

A recent study explored a range of indices of performance and its stability in four-finger tasks across the two hands (De Freitas et al., 2019). Earlier studies of such tasks focused on differences between the dominant and nondominant hands and reported advantage of the nondominant hand in steady force production tasks, while the dominant hand had an advantage in quick force production tasks

(Zhang et al., 2006; Park et al. 2012; Jo et al., 2015). These findings have been discussed as corroborating the dynamical dominance hypothesis (Sainburg, 2002, 2005), which assumes specialization of the dominant and nondominant cortical hemispheres for the control of fast actions and steady states, respectively. The aforementioned study by De Freitas and colleagues confirmed hand differences. It also discovered correlations between a number of indices of synergic control measured in the two hands across participants. In particular, there were no strong between-hand correlations for variables related to task performance, such as indices of performance accuracy. In contrast, significant correlations were seen in "hidden variables" that had no direct effect on performance (such as V_{UCM} and the index of motor equivalent motion) but reflected subject's preference with respect to stability and optimality of performance of the involved abundant systems, i.e., the four fingers. In other words, if a person showed high stability-related variables in one of the hands (e.g., high V_{UCM} suggesting low adherence to an optimality criterion), comparably high stability-related variables were seen in the other hand.

4.4 Activity of neuronal populations and action stability

The relations between activation of different brain areas and human body functions, including motor action characteristics, have been debated literally for centuries. In 1930s, Bernstein wrote a book "Contemporary Studies on the Physiology of the Neural Process" where he reviewed the state of the art of the field, argued with the then-dominant views of Pavlov and his school, and made a number of very bold statements. The book was ultimately published in Russian about 70 years later (Bernstein, 2003), nearly 40 years after Bernstein's death, and is still unavailable for the English-speaking community. In that book, Bernstein reviews the two dominant views on the relation between functions and their neural localization in the brain. One of the views was that any function had a dedicated neural substrate, which was responsible and specialized for that function. The other view was that any neural structure could contribute successfully to any function. Bernstein concluded that both extreme views were wrong: A number of neural structures contributed to any given function, but not each and every neural structure was equally suited to contribute to each and every function. Since the book was written, this conclusion has been supported by numerous studies, including clinical observations of specific functional deficits after injuries to specific brain areas and also widespread processes of neural plasticity contributing to functional recovery after a neurological injury, for example, in stroke survivors.

Most early studies of links between neuronal activation and characteristics of motor action in humans were focused on the cortical motor areas, which are relatively easily accessible for electrical stimulation. These areas are also the source of one of the major descending tracts, the corticospinal tract. Classical studies of the effects of electrical stimulation of cortical neurons demonstrated a somatotopic organization of the motor cortex (a distorted human body "drawn" on the cortical surface, Penfield and

Rasmussen, 1950) and correlations between stimulation strength and such action characteristics as muscle activation level and muscle force (Evarts, 1968; Asanuma, 1973). Following these classical studies, for some time, the view dominated that the motor cortex prescribed patterns of muscle activation and/or forces.

A groundbreaking series of papers were published by the group of Apostolos Georgopoulos in the 1980s (Georgopoulos et al., 1982, 1986; 1989; Georgopoulos, 1986). In those studies, a step was made toward linking activity of individual cortical neurons to activation patterns within populations of such cells, which were then compared to characteristics of movements performed by monkeys. The monkeys were implanted with arrays of small electrodes in the primary cortical motor area, and activity of individual neurons was recorded. When the monkeys performed a learned task of moving to targets arranged along a circle from a fixed initial hand position in the center of that circle, each neuron showed modulation of its activity with target location. This modulation typically followed a cosine function with a peak at some preferred target location, e.g., α_{PR} in Fig. 4.14A, and a drop with angular deviation from the preferred direction when the monkey moved to other targets. If a monkey moved away from the preferred direction of a neuron, activity of that neuron could show a drop from its baseline level. Such cosine functions were computed for each of the individual neurons. Then, when the monkey moved to a specific target, the activity of all the neurons was recorded, and each neuron was assigned a vector pointing in its preferred direction with the length equal to the number of action potentials this neuron generated within a time interval about the movement initiation. When all such vectors were summed up, the resultant vector across the population pointed very close to the direction of the target (α_{POP} in Fig. 4.14B).

Later, this method has been used to explore activation of neurons in other brain areas including other cortical areas, as well as other brain structures such as the

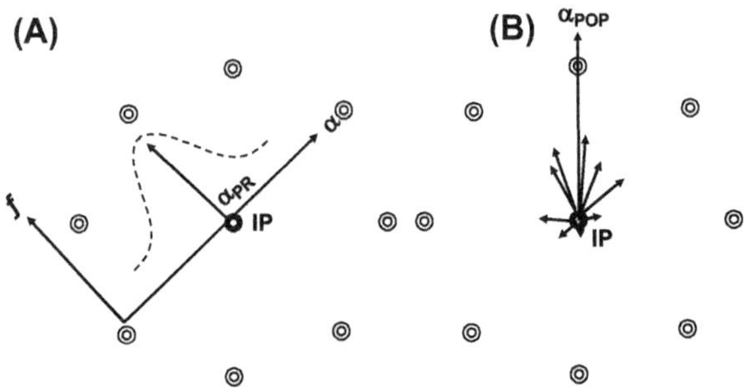

Figure 4.14 An illustration of the experiments with monkeys moving to targets arranged along a circle starting from a fixed initial hand position (IP) in the center of that circle. (A): Cortical neurons showed modulation of their activity with target location similar to a cosine function with a peak at some preferred target location, e.g., α_{PR}. B: When the monkey moved to a specific target, the sum of all individual vectors pointing in the preferred direction for each neuron with the length proportional to the number of action potentials this neuron generated within a time interval about the movement initiation, pointed very close to the direction of the target (the population vector, α_{POP}).

cerebellum (Fortier et al., 1989). The overall result has been confirmed in most studies with more or less noise (i.e., activity unrelated to movement direction) in the neuronal activation patterns. Properties of neuronal population vectors have been shown to depend not only on movement direction but also on the initial arm posture (Scott and Kalaska, 1995) and the external force field (Arce et al., 2010).

The finding of neuronal populations encoding peripheral motor variables has been very much promising as the means of improving quality of life of persons with paralysis because of interrupted transmission of signals from motor cortical areas to the spinal structures and muscles. This is particularly relevant to conditions leading to progressive loss of the motor function such as amyotrophic lateral sclerosis (known also as the Lou Gehrig disease). This condition is characterized by progressive death of α-MNs (and also neurons in other parts of the central nervous system) leading to loss of the ability to activate muscles. In severe cases, it may lead to a condition addressed as being "locked in the box," when the person cannot perform any voluntary motor actions (reviewed in Birbaumer et al., 2012). In such conditions, using brain neuronal activity to decipher desired actions and implement them with robotic devices could potentially help the patients in everyday life. Recent studies on brain−computer interface have shown promising results (Helms Tillery et al., 2003; Lebedev and Nicolelis 2017; Wolpaw et al., 2018), although they are still far from substituting for the lost motor function.

Most of the mentioned studies, starting from the cited classical studies by Apostolos Georgopoulos, explored trends seen in averaged across multiple trials data. Earlier experiments considered the whole population of neurons as a single unit with the output correlating with mechanical action characteristics. Recently, correlations among individual neuronal activation patterns have been explored assuming that cortical neurons within a population form groups, *neuronal modes*, which tend to show correlated pattern of activity and may be specialized for recruitment of specific muscle groups (Holdefer and Miller, 2002). Relations between activation patterns in neuronal populations and muscle activation patterns have been explored assuming a level of hypothetical *hidden variables* and looking for such variables with the help of canonical correlations (Pandarinath et al., 2018). Another approach has addressed the relations between muscle activation patterns and activity of neurons in the motor cortex and came up with a conclusion that, despite similarities between those two groups of signals, the cortical neurons had a specific feature: As a population they avoided tangling, situations where similar activity patterns led to dissimilar future patterns (Russo et al., 2018).

One common assumption across all these studies has been that activity in the brain, in particular in motor cortex, prescribes in some way future patterns of muscle activation. This is a questionable assumption given the action of spinal reflexes, including stretch reflex, which contribute to muscle activation patterns and make them dependent on the external forces, which are commonly unpredictable. Recent studies using transcranial magnetic stimulation applied over the primary motor cortex in human subjects have shown that cortical excitability may change without a visible change in muscle activation levels, while changes in muscle activation may be accompanied by no detectable changes in the cortical excitability (Raptis et al., 2010;

Ilmane et al., 2013). These observations suggest that descending signals from cortex may encode variables that affect muscle activation levels only indirectly (discussed in more detail in Part 5 of the book).

Another major issue that remains underexplored and is directly related to the problem of movement stability is whether neuronal populations show stability in encoding peripheral variables. Activity of individual neurons is typically noisy, partly because of their threshold properties. The question is as follows: Are neurons or neuronal modes within a population organized to stabilize a vector of their combined activity expressed in variables salient for performance? Or, in other words, if one neuron (mode), by chance, shows a deviation in its activation that moves the population vector in one direction, i.e., introduces an error into performance, will other neurons (modes) show deviations from their average activity compensating for this error? This seems to be a very important question. Indeed, if neuronal population vectors are used in brain—computer interfaces to drive external devices (e.g., robots), an unstable output from a neuronal population would produce an equally unstable action by the controlled robotic device, which would not be functionally useful. Computational methods to address this question have been developed, in particular in studies of multimuscle synergies. So, it is only a matter of time when stability of neuronal population outputs will be quantified in experiments and possibly optimized for clinical applications.

In the aforementioned book, Bernstein (2003) argued against the dominant view that the cortex was the ultimate controller of motor actions. He wrote: "No area of the cortex can currently be viewed as the origin or the final destination of a neural process … Every area and every layer of the cortex represent only transit points of the neural process (p. 326)." In a later classical paper (Bernstein, 1935), he wrote: "In the higher motor centers of the brain (very probably in the cortex of the large hemispheres) one can find a localized reflection of a projection of the external space in a form in which the subject perceives the external space with motor means" (cited after Bongaardt, 2001). This statement suggests that the motor cortex output encodes not mechanical variables associated with a planned action but neural variables reflecting perception by the subject of the task and external space expressed in movement-related variables. The last expression is directly linked to ideas within the field of ecological psychology developed about half-a-century later by Gibson (1979) and to issues of action-perception coupling addressed in Part 7.

Most of the reviewed studies of action stability explored stabilization of salient performance variables organized within multidimensional spaces of elemental variables measured as the outputs of elements at the selected level of analysis (such as joint rotations, digit forces, and muscle group activations). A recent shift to studies of action stability within spaces of hypothetical control variables is promising (Ambike et al. 2015; Reschechtko and Latash, 2017, 2018; see Section 5.6), but links to activation of neuronal populations remain elusive, although promising results have been obtained in studies that tried to link neuronal activity in the primary motor cortex to control variables within the EP hypothesis and its recent developments (Feldman, 2015), which is going to be described in the following part of the book.

4.5 Dealing with strong perturbations

Until now, we have been addressing local stability of actions, i.e., stability in cases of relatively mild perturbations associated, in particular, with spontaneous changes in intrinsic body states and natural variability of external forces in otherwise predictable conditions. Examples of such perturbations include stepping on a small pebble while walking, lifting an object that is slightly heavier or lighter than expected, taking a sip from a mug without looking at the level of liquid, etc. Sometimes, however, we have to deal with unexpected strong perturbations acting on the body that overpower the inherent dynamical stability of the ongoing action and require corrective sensory-based feedback reactions.

Imagine a walking person who suddenly steps on a slippery surface, for example, a patch of ice, and slips. The slip is typically followed by a very complex pattern of movements of all body parts including the arms, sometimes with a few clearly seen phases. In a large percentage of cases, the person restores balance and does not fall, but sometimes a slip does result in a fall. Each time a slip occurs, the movement pattern looks very complex and irreproducible, even unique. Such patterns are not seen in toddlers who simply flop on the ground; they are also not seen in older persons who sometimes fall with minimal corrective actions (and disastrous consequences). Success in slip recovery is much lower in persons brought up in warm climate where roads are free of snow and ice at all times. These general observations suggest that the corrective movement patterns represent results of development and learning, and they depend on proper functioning of the neuromotor system.

A particular approach to ensuring stability in cases of such strong perturbations has been suggested recently (Akulin et al., 2019) based on the general assumption that such highly variable patterns emerge as a result of a relatively simple algorithm applied to cases with varying initial conditions at the slip (or another cause of a major perturbation). Note that speed of reaction is essential to prevent loss of stability in such cases. So, no processes that require long central processing times can be used. While the brain is sometimes considered as an omnipotent processing machine that can solve problems within milliseconds, this is likely an exaggeration. As mentioned in another part of this book, the large difference between the simple reaction time (100–150 ms) and two-choice reaction time (>200 ms) suggests that making a simple decision of moving an effector in one of the two directions is rather time-costly. Hence, the approach advocated by the authors of the mentioned paper is that the central nervous system uses a simple set of rules without trying to find an optimal solution for the problem of recovering stability.

This approach may be seen as an extension of the more general idea that biological systems are reasonably sloppy (Latash, 2008; Loeb, 2012). This means that biological systems do not solve problems exactly, but rather use simple rules that produce solutions that are *good enough* (e.g., successful most of the time). Such rules may fail to solve specific problems and then healthy people make mistakes, fall, mishandle objects, spill coffee, etc. But they are successful in a large proportion of everyday situations. To implement this idea, the authors suggested a rather simple algorithm, which they addressed as the "act on the most nimble" rule (the AMN rule).

The AMN rule assumes that the neural controller acts in a space of *control coordinates* representing combinations of elemental variables that allow using local control. This means that the controller can act on one of those variables and expect changes in that very variable, not in other variables. The AMN rule states that changes in control coordinates occur for the most unstable variable first, i.e., in a variable that shows the fastest deviations from its expected trajectory. If this first corrective action is unsuccessful, the controller acts along the second fastest coordinate, and so on until the action is stabilized.

As an illustration of the AMN rule, imagine that you have to take two dozen of preschool children for a walk. In such a group, there are typically obedient, well-behaving children and a few unruly ones who try to run in different directions at all times. How can you make sure that the group, as a whole, moves where you want it to move? Trying to prescribe each and every kid what to do is unrealistic. One strategy is to control the unruliest kid and thus minimize the disruption of the group he or she produces. This may work unless the second unruliest one takes over. Then, you switch to the second one, and so on, until the group acts as a whole in a desired way.

Related issues have been addressed in the field of control theory, in particular, optimal feedback control (Diedrichsen et al., 2010; Todorov and Jordan, 2002). There are, however, a few major differences between the two approaches:

First, the discussed approach does not involve computing a value of a cost function, only a simple rule; second, the approach is expected to be specific to problems of restoring stability of action and it does not address purposeful transitions to another state; and third, this approach assumes no knowledge of the central nervous system of mapping elemental variables onto performance variables. In other words, it does not assume that the controller is aware of the Jacobian matrix (**J**) of the system. This is arguably the most important difference from earlier approaches. It is intuitively attractive. For example, when we learn to use new tools, from a tennis racket to a car, we develop sets of rules leading to success without trying to compute all the new transformations from neural signals to muscle action and to motion of the racket or the car in space. While the approach based on the AMN rule is algorithmic, the authors assume that the rule reflects physical (including physiological) processes within the body and that the mathematical descriptions are reflections of those processes. No such computations are assumed to take place within the body.

The paper by Akulin and colleagues (2019) presents the results of a couple of experiments with multifinger force production. One of them showed that, indeed, during unexpected perturbations humans tend to act along the "most nimble" directions in the space of elemental variables. The other experiment showed a sequence of corrections in cases of unknown and unexpected changes in the task that resembled those computed in a model implementing the AMN rule (Fig. 4.15). In this study, the subjects were asked to produce target values of total finger force and total moment of force produced with respect to the midline between the middle and ring fingers with the help of visual feedback on the computer monitor.

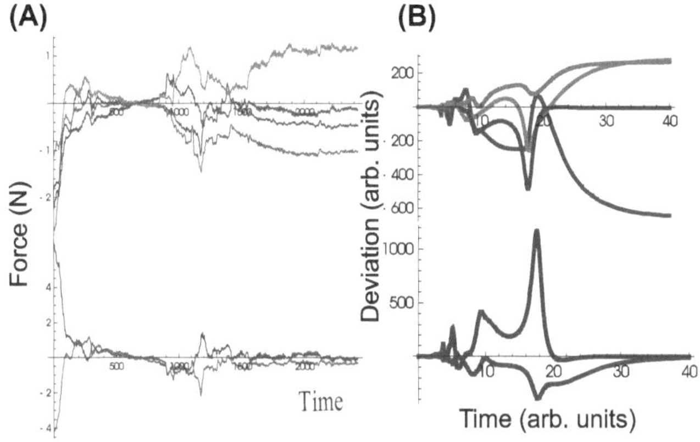

Figure 4.15 (A): The time profiles of four finger forces (top) and two task-related variables (bottom) when performance stability was lost because of a change in the Jacobian mapping finger forces on the task-related variables. (B): Results of a simulation of a system of four elements ("fingers") performing a two-dimensional task leading to successful stabilization of performance. The time profiles of finger forces are shown in the top panel, and task-related variables are shown in the bottom panel. Note the clear discrete episodes of corrections seen in both experiment and simulation.
Reproduced by permission from Akulin, VM., Carlier, F., Solnik, S., Latash, ML., 2019. Sloppy, but acceptable, control of biological movement: Algorithm-based stabilization of subspaces in abundant spaces. Journal of Human Kinetics (in press). (Copyright by DeGruyter).

Then, unexpectedly, the mapping of the finger forces to the two variables on the monitor changed. For example, one of the fingers could suddenly become "weak" or started to press upward. Spontaneous finger force changes led to movement of the cursor on the screen and when the subjects tried to bring the cursor back into the target, it suddenly started to move according to the new (unknown to the subject) rule. In about 80% of cases, subjects were able to restore stability of the cursor in the middle of the screen within 20–30 s. As shown in Fig. 4.15, there were clear episodes of corrective actions that looked like AMN-based actions.

Where does the AMN rule come from? We do not know (BMC!). Likely it can be discovered, refined, and optimized in the process of practice. The process of discovering acceptable solutions to the task of stabilizing a performance variable requires, as the first step, finding a set of adequate control coordinates. This process may be crucial, in particular, for the development of skills, which require high stability of salient performance variables during frequent and unexpected changes in the external forces acting on the body. Typical examples may include downhill skiing, snowboarding, gymnastics, and other sports characterized by large, quickly changing forces between the athlete's body and the environment. It is feasible that the central nervous system discovers several sets of control coordinates, which have to switch at different phases of such complex movement. Discovering precise switching times may represent another step of learning a complex skill.

On the other end of the spectrum of human motor abilities, the ability to discover adequate sets of control coordinates to be used in the main algorithm and the ability to switch from one set of control coordinates to another can be violated in various motor disorders. These processes may suffer from inadequate sensory input, for example, in patients with large-fiber peripheral neuropathy (so-called "deafferented persons" Sainburg et al., 1993; Yousif et al., 2015). This condition may not allow forming adequate sets of control coordinates and providing sensory information on changes in elemental variables to be used in implementation of the AMN rule. It may also impair the ability to time switches among sets of control coordinates appropriately. The mentioned mechanisms may also suffer from an injury to neural structures, in particular components of neural loops via the basal ganglia and the cerebellum to the cortex. Note that loss of movement stability and ability to learn stable performance in novel conditions is a sign of neurological disorders involving the cerebellum and the basal ganglia (reviewed in Llinas and Welsh 1993; Celnik, 2005; Latash and Huang, 2015).

Control with referent coordinates

5

Arguably, the first hypothesis in the field of biological movements based on parametric control, the equilibrium-point (EP) hypothesis, was suggested over 50 years ago (Feldman, 1966). The fate of this hypothesis is very unusual. Over the past half a century, it has neither been disproven nor accepted by most of the colleagues/researchers. There were times then the EP hypothesis was ignored and times when it was fashionable to disprove it. It has been declared dead multiple times based on both theoretical considerations and experimental observations (Gomi and Kawato, 1996; Gottlieb, 1998; Popescu and Rymer, 2000; Hinder and Milner, 2003; Shadmehr and Wise, 2005). However, later, all those attempts have been shown to be invalid because they were based on misunderstandings of one or more of the basic assumptions and central ideas of the EP hypothesis (Feldman and Latash, 2005; Feldman, 2015).

Originally, the EP hypothesis was suggested as a concise description of a body of experimental observations in both animal and human studies (Matthews, 1959; Asatryan and Feldman, 1965). All those studies showed that the dependence of active muscle force (F_A) on muscle length (L), both measured at steady states, could be described as a monotonically increasing function with a parameter (λ) corresponding to the length value when the muscle showed first signs of activation: $F_A = f(L - \lambda)$.

Fig. 5.1 illustrates schematically the main results of an experimental study using human subjects with the $F_A(L)$ functions for a pair of opposing muscles. In this figure, two different variables are used, joint angle and moment of force, which are more adequate to describe joint rotational actions. One of the muscles is assumed to produce positive moment values (agonist), whereas the other produces negative moment values (antagonist).

The functions illustrated in Fig. 5.1 were obtained using the procedure of partial unloading of the elbow joint under the instruction to the subject "do not intervene voluntarily with arm motion induced by changes in the external load" Following this procedure, a set of equilibrium states of the "joint + reflexes + load" system was obtained and then the EPs were interpolated to approximate the $F_A = f(L - \lambda)$ function (I continue using the F and L pairs of variables to avoid proliferation of symbols). In another series, the same subject was asked to hold a different load in the same initial joint position and then, once again, "not to interfere" when the load was changed. Let us pay attention to two important facts in these results. First, when a person tries to keeps the "command" constant, equilibrium states follow a curve with an invariant intercept with the length axis, λ. Second, when the same subject changes the initial state (changes the initial "command"), λ changes and then remains at the new value during the unloading procedure leading to a shift of the $F_A(L)$ characteristic. In this context, the word "command" in quotation marks means the so far undefined

Physics of Biological Action and Perception. https://doi.org/10.1016/B978-0-12-819284-9.00005-X

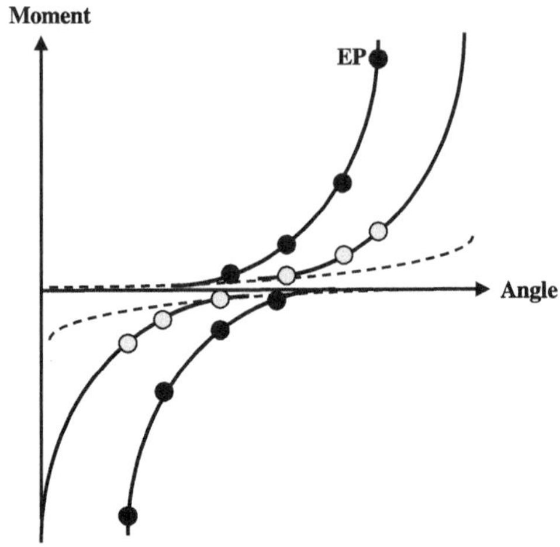

Figure 5.1 A schematic illustration of the dependences of muscle moment of force on joint angle for a pair of opposing muscles; note that one of the muscles produces negative moment values. The *dashed lines* show the dependences for quiescent muscles. The *solid lines* are interpolations of the equilibrium points (EP) when the subject was asked to hold a joint position against a constant load. Then the load was decreased under the instruction to the subject "do not intervene voluntarily with arm motion induced by changes in the external load" Note that the dependence of moment on angle remained the same as long as the subject did not interfere; it changed when a different load was held at the same initial joint position (compare the *black and gray circles*).

combination of descending signals, potentially from different brain structures, associated with performing this motor task.

Can a person follow the instruction "to keep the command constant"? Maybe, there are unintentional adjustments in the descending signals from the brain to the unloading procedure, leading to the observed invariance of λ. This is an important possibility. A few observations suggest that the results are not artifacts of a specific pattern of unintentional interference by the subjects. First, similar sets of $F_A(L)$ curves were obtained in experiments with animal preparations (e.g., so-called, *decerebrate preparations*) when the large hemispheres and some other structures of the brain were surgically separated from the spinal cord and the brain stem (Matthews, 1959; Feldman and Orlovsky, 1972). It is fair to assume that these preparations were unable to interfere voluntarily because the main brain structures associated with voluntary action were unable to communicate with the spinal cord. In another study, human subjects were instructed to react voluntarily to changes in the external load: Specifically, they were instructed to return to the initial posture when the external load changed (Latash, 1994). Behavior under this instruction was much less reproducible as compared with the behavior under the "do not intervene voluntarily" instruction.

In what sense are the $F_A(L)$ dependences invariant? This is another good question. Indeed, effects of hysteresis have been known in the field of muscle behavior for a long time (e.g., Partridge, 1965). So, loading and unloading under the same instruction may be expected to lead to different equilibrium states, i.e., different joint angles when the load comes to the same final magnitude (confirmed by Gottlieb and Agarwal, 1986, 1988). This is not, however, a crucial fact. Indeed, one can control movements with more or less predictable (more or less invariant) $F_A(L)$ characteristics. The issue is not whether such characteristics are 100% reliable and predictable but whether they form a set of tools that can be used by the brain to control movements. We will return to this issue later.

The idea of controlling movements by changes in parameters of muscle force–length characteristics had not been 100% novel by the time the EP hypothesis was suggested. The idea of control with parameters of muscle reflexes dates back to classical studies of Charles Sherrington who also introduced the notion of tonic stretch reflex as a mechanism underlying the $F_A(L)$ dependence (Liddell and Sherrington, 1924). In his main book (which is still unavailable in English!), Bernstein (1947) suggested that the brain could send descending signals shifting the muscle $F_A(L)$ characteristics parallel to the length axis. In one of the figures in that book, Bernstein drew linear $F_A(L)$ characteristics, but this does seem to represent a crucial difference from the characteristics found by Anatol Feldman. Bernstein, however, did not develop this idea for the neural control of movements.

In the early 1950s, a *servo-model* was introduced by Merton (1953), which shared certain important features with the EP hypothesis. Merton suggested that the brain encoded desired postures with signals to gamma-motoneurons, small neurons that send their axons to muscle fibers within structures called muscle spindles. These muscle fibers do not connect to tendons and do not produce joint moment of force. Their activation leads to changes in the sensitivity of small sensory endings located on the fibers to muscle length and velocity; these sensory endings produce excitatory effects on alpha-motoneurons innervating the muscle. Based on these facts, Merton suggested that a signal to gamma-motoneurons encoded a desired value of muscle length that could be achieved independently of the external load. The scheme, illustrated in Fig. 5.2,

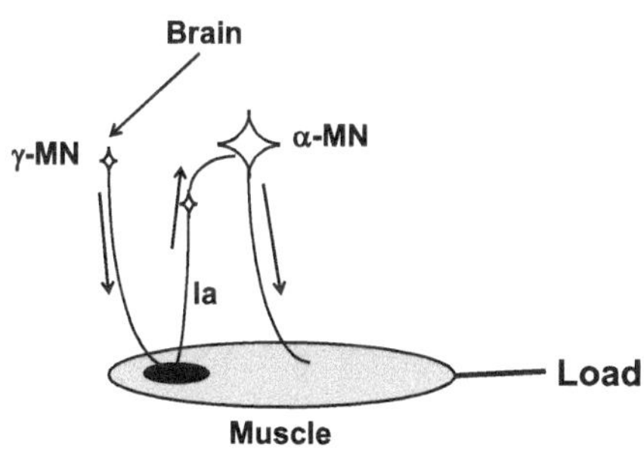

Figure 5.2 A scheme of Merton's servo-hypothesis. The brain is assumed to encode desired postures (muscle length values) with signals to gamma-motoneurons (γ-MN), small neurons that send their axons to muscle fibers within structures called muscle spindles (intrafusal fibers). Their activation leads to changes in the sensitivity of small sensory endings located on the intrafusal fibers to muscle length and velocity. A signal to gamma-motoneurons leads to activation of alpha-motoneurons (α-MN) via the stretch reflex loop, which leads to changes in muscle activation and its length changes until the system comes to an equilibrium at the desired muscle length value.

included the stretch reflex as the main mechanism, ensuring changes in muscle activation as long as its length differed from the value preset by the signal to gamma motoneurons. In other words, the stretch reflex was assumed to act like a perfect feedback servo moving the system to a desired position.

Note the following important features of the servo-hypothesis. First, it assumes that movements are controlled not by setting time profiles of muscle forces or muscle activations (both Bernstein and Merton were convinced that this was impossible, given the design of the human body) but by setting parameters of the stretch reflex leading to shifts of the $F_A(L)$ characteristics. It also assumes that the system "muscle + reflexes + load" comes to equilibrium states defined by the stretch reflex parameters. These two features make the servo-model similar to the EP hypothesis.

The next three features, however, make the servo-hypothesis and EP hypothesis different both quantitatively and qualitatively. First, the servo-model assumes very high gain in the stretch reflex loop; this assumption was later falsified in experiments (Vallbo, 1970). Second, the servo-model assumes that the neural command terminates on gamma-motoneurons, small cells with thin axons and relatively slow conduction velocity of action potentials, while activation of alpha-motoneurons comes at a delay defined primarily by the limited speed of transmission of action potentials along neural fibers. This prediction was also shown to be wrong: Both types of motoneurons show changes in their activation level nearly simultaneously, a phenomenon addressed as alpha-gamma coactivation (reviewed in Granit, 1970). Third, the servo-model can account for movements but not for force production in isometric conditions. Indeed, the control with nearly vertical $F_A(L)$ characteristics (illustrated in Fig. 5.3) makes force changes very large in response to tiny changes in the control variables (location of the characteristic). In other words, the system becomes unstable in kinetic tasks, a prediction not supported by experiments and everyday experience: For example, people can grip an object and keep it gripped in the hand for a long time with only small variations in the gripping force.

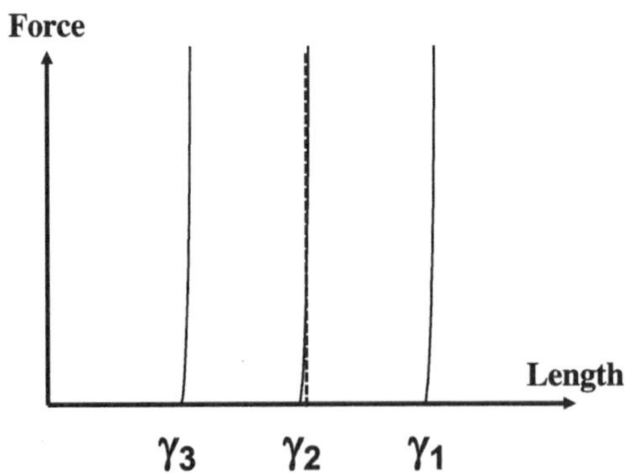

Figure 5.3 An illustration of Merton's servo-hypothesis with force−length muscle characteristics. Such characteristics are nearly vertical; this allows encoding muscle length independently of possible variations in the external load with signals to gamma-motoneurons (γ). Note that in isometric conditions (*vertical dashed line*), this system cannot produce accurate force magnitudes.

To summarize, the EP hypothesis was built on the strong foundation of both theoretical concepts and experimental material. Before moving to specific features of the neural control of different effectors, let me summarize the main distinguishing features of the EP hypothesis:

It is based on a well-established fact, a dependence of active muscle force on muscle length, which may be viewed as a biological law of nature (see Section 2.2.1);

It suggests a particular realization of the principle of parametric control, namely a change in threshold of the stretch reflex, λ, and

It does not assume computational processes within the body (including the brain), only physical/physiological processes that form a set of tools used by the brain to produce movements.

5.1 The equilibrium-point hypothesis: the muscle level

The EP hypothesis has been most thoroughly developed for the control of a single muscle (reviewed in Feldman, 1986; Feldman and Levin, 1995). The control variable (λ) has been associated with a well-defined physiological variable, depolarization of the membrane of alpha-motoneurons. Note that, in the original papers, λ was associated with threshold muscle length when the muscle shows first signs of activation during very slow stretch—threshold of the tonic stretch reflex. This variable is measured in units of length, meters. Membrane depolarization is measured in volts (or, better, millivolts), units that are adequate for neural processes. So, the EP hypothesis uses the stretch reflex mechanism to link two "worlds," the world of external mechanics and the world of intrinsic neural processes via a variable that can be expressed in both units. This process does not require an interpreter, for example, a computational device, which would compute how many meters correspond to one millivolt in each given situation.

This point deserves a more detailed explanation. Imagine an alpha-motoneuron that innervates a group of muscle fibers and also receives excitatory signals from length-sensitive sensory endings located in the same muscle (E1 in Fig. 5.4A). When the excitatory signals are relatively weak, they produce small, slow deviations of the membrane potential from its equilibrium value, which are compensated by the ion pump mechanisms that keep the equilibrium potential steady (see the thin dashed line in Fig. 5.4B). If the muscle is slowly stretched, the excitatory signals from the length-sensitive receptors increase in magnitude and, at some threshold value of muscle length, the compensatory mechanisms might be unable to compensate for the growth of the membrane potential, and the potential would grow toward the threshold for action potential generation (the thin solid lines in Fig. 5.4B). At this value of muscle length, first signs of muscle activation can be seen. Note that, at muscle length values close to the threshold for activation, the growth of the membrane potential is slow and, correspondingly, the frequency of action potential generation is low (see the two small-font APs in Fig. 5.4B). After each action potential, the membrane is unexcitable for a very short time (referred to as absolute *refractory period*), and then the process repeats itself. Stretching the muscle leads to an increase in the rate of membrane potential deviation from the equilibrium potential and to an increase in the frequency of action potential generation (see the six large-font APs in Fig. 5.4B).

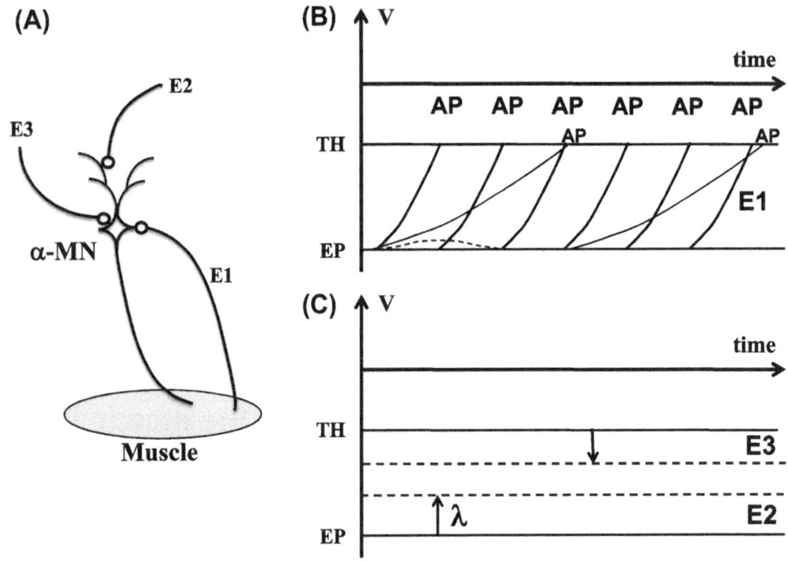

Figure 5.4 An illustration of the physiological mechanisms involved in the control of a single muscle. (A): An α-motoneuron innervating a group of muscle fibers receives excitatory signals from length-sensitive sensory endings located in the same muscle (E1) and from two sources within the central nervous system (E2 and E3). (B): Weak E1 produces small deviations of the membrane potential from its equilibrium value (EP), which do not reach the threshold (TH) for action potential (AP) generation (thin, *dashed line*). At some, longer value of muscle length, first signs of muscle activation can be seen (*thin lines* and small-font AP). At longer values of muscle length, E1 is stronger and leads to high-frequency APs (*thicker lines* and large-font APs). (C): E2 leads to subthreshold depolarization of the membrane resulting in a change in stretch reflex threshold (λ). E3 leads to a drop in the TH magnitude.

Consider now two more inputs into the same alpha-motoneuron. One of them (E2 in panels A and C of Fig. 5.4) represents descending muscle control and produces steady depolarization of the membrane potential; the depolarization effectively decreases the distance to the threshold for action potential generation (dashed lines in Fig. 5.4C). This helps the E1 input to the alpha-motoneuron reach the threshold earlier; for smaller values of the muscle length, and for any given muscle length longer than the threshold value, the frequency of action potential generation increases. When this input is zero, the muscle is relaxed over its anatomical range: Indeed, humans are able to relax healthy muscles at any anatomically possible value of their length. An increase in E2 allows the muscle to show signs of activation at some, relatively long, length. Further increase in E2 allows the muscle to show activation at shorter length values. This illustration shows once again that descending muscle control can be expressed in units of membrane depolarization (mV) as well as in units of muscle length (m) corresponding to the threshold value for muscle activation.

The third input into the motoneuron (E3 in panels A and C of Fig. 5.4) has drawn attention relatively recently. It reflects the phenomenon of *persistent inward currents*

(PIC; reviewed in Heckman et al., 2005, 2008), a depolarizing current produced via voltage-sensitive Ca^{++} channels that do not show inactivation, i.e., a drop in the permeability to ions following a brief episode of the channel being open (channel inactivation is typical, for example, for Na^+ channels). This lack of inactivation justifies the term *persistent*. PICs are typical of processes in dendrites of motoneurons. Effectively, PICs may be seen as the means of reducing the threshold for action potential generation and, hence, reducing the distance from the equilibrium potential to this threshold. So, PICs seem to work synergistically with the more traditional descending inputs (E2). On the other hand, once active, PICs tend to sharply limit efficacy of additional synaptic input; so there seems to be no direct summation of the effects of E2 and E3, as it may seem from the simplistic scheme in Fig. 5.4.

Two types of movements are possible within the EP hypothesis, *passive* and *active* (also addressed as *voluntary*). A change in the external load without a change in λ leads to a passive movement, and the system comes to rest at a different point of the $F_A(L)$ characteristic defined by λ (a change from load L1 to load L2 leads to a shift in the EP from EP_0 to EP_1 in Fig. 5.5). Note that passive movements are associated with changes in muscle activation level due to the stretch reflex action.

Active movements are associated with changes in λ. Then, the original $F_A(L)$ characteristic disappears and a new one emerges. The original EP vanishes, and the system moves to a new EP. This motion is not predefined by the central control process because it also depends on the external load. Indeed, as illustrated in Fig. 5.5, the same shift of λ (from λ_1 to λ_2) can lead to a force change (in isometric conditions, i.e., when the muscle cannot change its length, to EP_2), to a change in muscle length

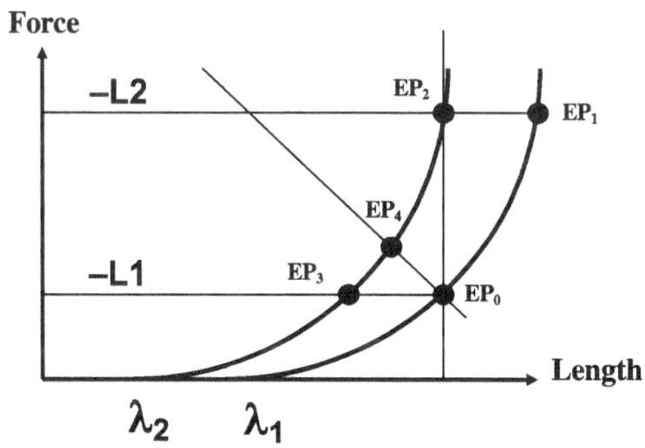

Figure 5.5 According to the equilibrium-point (EP) hypothesis, the central nervous system controls each muscle by specifying its threshold of the stretch reflex (λ). For a given external load (e.g., L1), the muscle comes to equilibrium at a state when it produces force equal in magnitude to the load (EP_0). A change in the load (e.g., to L2) leads to motion along the force–length characteristic specified by λ (passive movement to EP_1). A shift in λ (e.g., from λ_1 to λ_2) can lead to different mechanical consequences depending on the external load characteristic (shown by *thin lines*). In isometric conditions, it leads to force change (EP_2); in isotonic conditions—to length change (EP_3); and in more natural conditions—to a change in both force and length (EP_4).

(in isotonic conditions, i.e., when the external load remains constant, to EP_3), or to both (for a more realistic load that changes during the movement, to EP_4).

As illustrated in Fig. 5.5, within the EP hypothesis, movement and force generation are different peripheral outcomes of basically the same control process, a shift of λ, and the difference in motor behavior is due to different external forces acting on the muscle.

Of course, this is a very much simplified description. In fact, each muscle consists of many motor units, i.e., groups of muscle fibers that receive excitation from branches of a single neural fiber (the axon) originating from a single alpha-motoneuron. Each motor unit may be viewed as a "mini-muscle" with its own stretch reflex, leading to changes in the frequency of action potentials generated by the alpha-motoneuron in response to muscle stretch. Then, each "mini-muscle" has its own control signal, "mini-λ" associated with depolarization of the membrane of the corresponding alpha-motoneuron. The overall muscle λ is the minimal of the motor unit "mini-λs," and the muscle $F_A(L)$ characteristic is defined by both an increase in the firing frequency of already recruited motor units and recruitment of new motor units.

As mentioned earlier, the stretch reflex may be viewed as a mechanism uniting individual motor units into a synergy stabilizing the overall equilibrium state of the muscle acting against an external load (Fig. 4.12). For a constant λ, effects of spontaneous changes in activation levels of individual motoneurons are compensated, at least partly, by adjustments in the activation levels of other motoneurons of the pool mediated by the stretch reflex. The last statement remains a prediction, which has not yet been tested experimentally.

A number of muscles in the human body are traditionally viewed as a combination of muscle compartments, defined as groups of motor units that produce similar mechanical actions, while the actions differ across the compartments. For example, movements of the human fingers are produced, in particular, by the action of two extrinsic flexor muscles, flexor digitorum profundus (FDP) and flexor digitorum superficialis (FDS). Each of these muscles has its belly in the forearm and is connected by four tendons to the individual fingers. Although both FDP and FDS are described in anatomy books as individual muscles, physiologically and biomechanically they are commonly viewed as consisting of four compartments each. Each compartment acts primarily on one of the four tendons and causes action by one of the fingers. There is a degree of interaction among the compartments, and this interaction is viewed as a contributor to finger interdependence or lack of individuation (reviewed in Schieber and Santello, 2004). Indeed, if you try to wiggle one of the fingers, e.g., the ring finger, the other fingers of the hand will also wiggle, although over smaller amplitudes. Observations of professional pianists have shown that massive practice can lead to much larger finger independence (Slobounov et al., 2002), likely mediated by changes at a neural level without a major change in the muscle anatomy.

Overall, selecting muscle as the main level of analysis in the original versions of the EP hypothesis was rather arbitrary, maybe due to tradition. As illustrated in the former few paragraphs, one can move into the muscle and consider it as a constellation of compartments or of motor units. One can also move to more functional units of motor behavior such as joints that are crossed by multiple muscles.

5.2 Control at the joint level

Arguably, the lowest functional level of movement analysis is that of individual joint rotations. Indeed, humans understand and can follow the instruction "move such-and-such joint in a such-and-such direction" (for example, flex the elbow) but not the instruction "activate such-and-such muscle" (for example, activate the long head of triceps). Although all joints are crossed by multiple muscles—sometimes muscles cross more than one joint (so-called biarticular and multiarticular muscles)—let us consider a cartoon joint with a single axis of rotation that is crossed by only two muscles acting against each other. To simplify description, let us also view the two muscles as having similar mechanical and reflex properties resulting in nearly identical shapes of the $F_A(L)$ characteristics. Of course, to shift from analysis of muscle action to joint action, we have to shift from the force–length variables to those describing rotational action, moment of force or torque (T_A), and angle (α).

Fig. 5.6 illustrates the $T_A(\alpha)$ characteristics of the two assumed muscles, agonist and antagonist. One of the muscles (the agonist) produces positive torque values, whereas the other muscle (the antagonist) produces negative torques. Correspondingly, the $T_A(\alpha)$ characteristics of the antagonist are turned upside down. Joint behavior will be defined by the algebraic sum of the two characteristics (the thick, dashed line in Fig. 5.6), which, under the assumption of identical muscle properties, is straight within the range where both muscles are active, the so-called *coactivation zone*.

The control of a joint illustrated in Fig. 5.6 can be described with time changes in two variables, $\lambda_{AG}(t)$ and $\lambda_{ANT}(t)$, where the subscripts refer to the agonist and antagonist muscles. Alternatively, two other variables can be used, addressed as

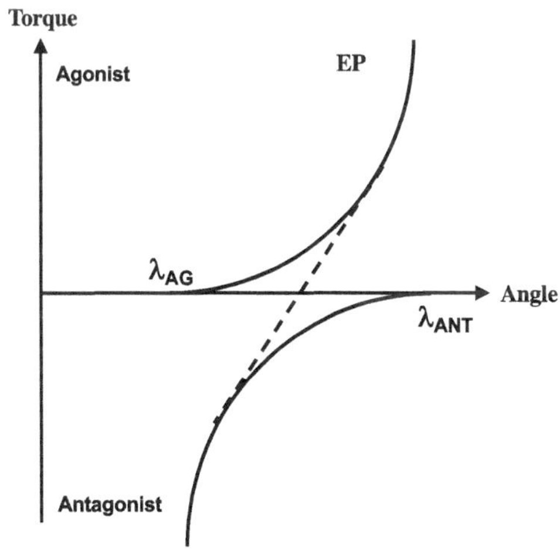

Figure 5.6 Mechanical behavior of a joint is defined by the algebraic sum of the agonist (the muscle producing positive torque values) and antagonist (the muscle producing negative torque values) muscle characteristics. The joint characteristic is close to being straight within the range where both muscles are active under the (unrealistic) assumption of identical muscle properties.

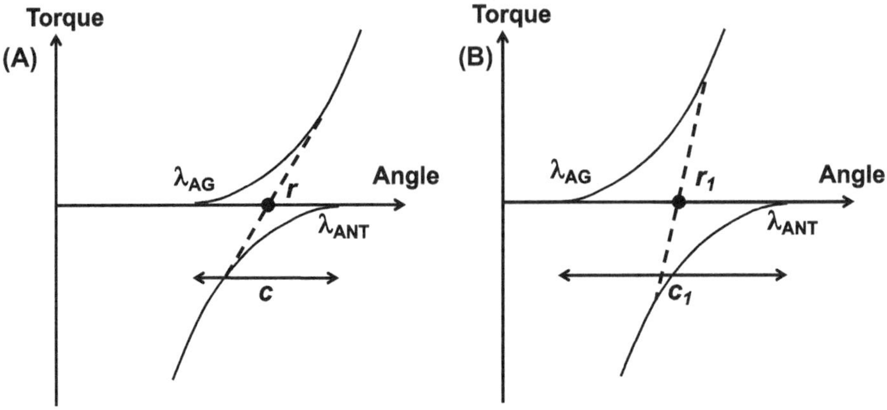

Figure 5.7 (A): The control of a joint crossed by two muscles, an agonist and an antagonist, can be described with two variables, $\lambda_{AG}(t)$ and $\lambda_{ANT}(t)$. Alternatively, two other variables can be used addressed as reciprocal command (r) and coactivation command (c). (B): A change in r (to r_1) effectively shifts the joint characteristic (*dashed line*) along the length axis, whereas a change in c (to c_1) effectively changes the apparent stiffness of the joint (the slope of the joint characteristic).

reciprocal command (r-command) and *coactivation command* (c-command) (Fig. 5.7; see also Feldman, 1980). This new pair of variables maps well on the intuitively understood two qualitatively different actions by a joint: One can move the joint (assuming that the movement is not blocked; if joint movement is blocked, the same change in the r-command would lead to a change in the net joint torque) and also cocontract the muscles and increase the joint's *apparent stiffness* (Latash and Zatsiorsky, 1993) without moving it. Formally, the two new variables have been defined as the midpoint between λ_{AG} and λ_{ANT} ($r = 0.5 \cdot [\lambda_{AG} + \lambda_{ANT}]$) and the range between λ_{AG} and λ_{ANT} ($c = \lambda_{AG} - \lambda_{ANT}$) where both muscles show nonzero activation levels (Fig. 5.7A). Note that both r- and c-commands are expressed in spatial units, although the latter effectively translates into changes in the joint apparent stiffness: A change in the c-command leads to a change in the slope of the joint $T_A(\alpha)$ characteristic, whereas a change in the r-command leads to translation of the characteristic along the joint axis as illustrated in Fig. 5.7B.

The definition of the c-command illustrated in Fig. 5.6 formally allows its values to be both positive, when the activation ranges of the two muscles partially overlap (shown in the top panel of Fig. 5.8), and negative, when there is a joint angle range where both muscles are quiescent (the bottom panel of Fig. 5.8). Negative coactivation does not seem to make much sense. Indeed, most studies of muscle coactivation define it using muscle activation indices that produce only nonnegative values (reviewed in Smith, 1981). When one of the two opposing muscles is quiescent, coactivation is postulated to be zero. It is definitely zero, when both muscles show no identifiable activity. So, the definition of the c-command, illustrated in Figs. 5.5 and 5.6, shows that, despite its formal name (coactivation command), it has no one-to-one correspondence with the phenomenon of muscle coactivation as reflected in levels of muscle activation.

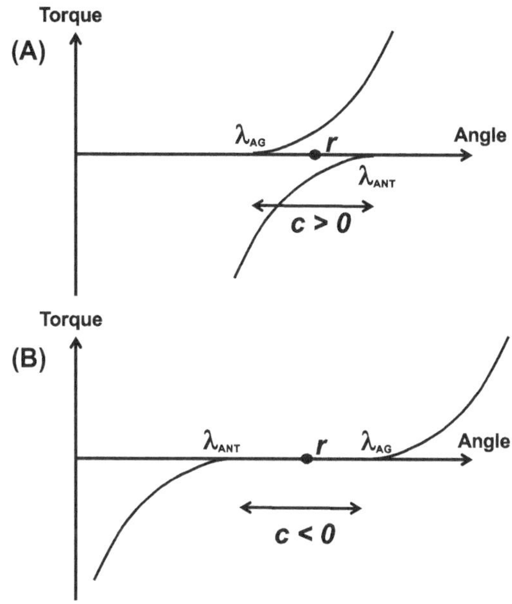

Figure 5.8 The definition of the c-command formally allows its values to be both positive ((A), when the activation ranges of the two muscles partially overlap) and negative ((B), when there is a joint angle range where both muscles are quiescent).

The c-command can be assigned functional significance that goes beyond changing the joint apparent stiffness. In particular, it is very important to consider that sensory endings contributing to the muscle $F_A(L)$ characteristic are velocity sensitive. Given this factor, a transient increase in the c-command in the middle of a fast action is expected to contribute to forces produced by the agonist during the acceleration phase and to forces produced by the antagonist during the deceleration phase (reviewed in Feldman, 2015), thus making movements faster while, at the same time, avoiding excessive oscillations at the end of the movement. Note that a transient increase in the c-command has been reported in several experimental studies, and the magnitude of the increase shows the expected scaling with movement velocity (Latash and Gottlieb, 1991, 1992).

There is one more potentially important role of the c-command related to the fact that the $\{r; c\}$ pair forms an abundant set of control variables with respect to action of a unidimensional effector (such as a joint with one rotational degree of freedom). This topic is considered in more detail later (see Sections 5.6.1–5.6.4). For now, imagine the task of producing a certain magnitude of joint torque in isometric conditions, i.e., when the joint acts against a stop (Fig. 5.9). Of course, nobody is perfect, so the subjects of this mental experiment are expected to be within a certain (small) error margin centered about the required torque magnitude.

Note that any torque magnitude can be produced with an infinite number of $\{r; c\}$ combinations (three such pairs are illustrated in Fig. 5.9) reflected in pairs of joint apparent stiffness (k) and referent coordinate (RC). For simplicity, assume a linear

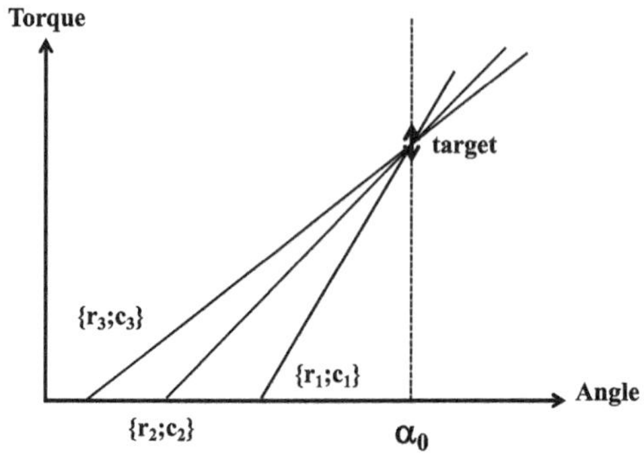

Figure 5.9 The task of producing a magnitude of joint torque into a small target in isometric conditions (joint angle is always α_0) can be performed with different $\{r; c\}$ combinations (three such pairs are illustrated).

system (this assumption is not crucial). Then, joint torque can be computed as $T = k \cdot (AC - RC)$, where AC is joint actual coordinate. Let us also select a system of coordinates where the actual joint coordinate corresponds to $AC = 0$; this step is also not crucial. Then, $T = -k \cdot RC$. This equation suggests that, for any given value of T, there exists a one-dimensional subspace within the two-dimensional $\{RC; k\}$ space where this value is not changing. This subspace, the uncontrolled manifold (UCM) for this task, is a hyperbolic line illustrated in Fig. 5.10. Several experiments (performed with a different effector, a finger rather than a joint) have documented very wide spreads of data points along such a UCM over a set of repetitive trials with the production of the same mechanical output (Ambike et al., 2015; Reschechtko and Latash, 2017). Note that this result is not obvious. The subjects were free to select a combination of the $\{r; c\}$ commands leading to the corresponding $\{RC; k\}$ values, e.g., an optimal combination according to some criterion, and reproduce it with

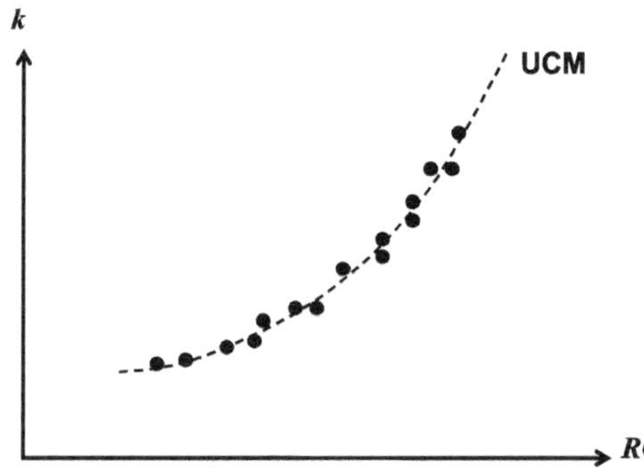

Figure 5.10 In the two-dimensional space of joint referent coordinate (RC) and apparent stiffness (k), the task of producing a magnitude of joint torque corresponds to a *hyperbolic line*, the uncontrolled manifold (UCM) for this task. Experiments show large variability of the data points across repetitive trials along the UCM with very small variability orthogonal to the UCM.

minimal inter-trial variability. This strategy would produce accurate performance without high variability along the UCM.

The aforementioned experimental findings (large variability of the {*RC*; *k*} values along the UCM) have been interpreted as a sign of a {*RC*; *k*} synergy organized at the level of neural control variables stabilizing the desired mechanical output. Consider now the situation of producing a certain torque value when the *c*-command is negative (as in the bottom panel of Fig. 5.6). Note that only one muscle, the agonist, contributes to the task as changes in λ_{ANT} lead to no mechanical effects. So, the abundant {*RC*; *k*} space becomes degenerate, one-dimensional, and there is no UCM corresponding to a desired torque value. This is expected to lead to lower stability of performance as any deviations of λ_{AG} lead to errors in torque with no chance to compensate for the errors by changing the other control variable.

This analysis suggests that agonist–antagonist coactivation may reflect a neural control strategy with the purpose to stabilize the ongoing action at a neural level of {*RC*; *k*} synergies (Latash, 2018b). This may be the cause of nonzero muscle coactivation, which is typical for most everyday actions. The level of coactivation increases in challenging conditions and in various populations with impaired motor function, from the healthy elderly to persons with atypical development and neurological disorders (Hammond et al., 1988; Aruin et al., 1996; Lee and Ashton-Miller 2011; Arias et al., 2012; Richards and Malouin, 2013; Kitatani et al., 2016). Tasks perceived as challenging are typically characterized by possible unexpected changes in external forces and/or intrinsic body states. Such unexpected changes may lead to movements when one of the opposing muscles exits the coactivation zone. As a result, the neural control can unexpectedly become degenerate, thus contributing to destabilization of the ongoing action. One of the feasible adaptive strategies is to increase the coactivation zone, i.e., increase the *c*-command. This would ensure that unexpected changes in the external and intrinsic states are less likely to move the system outside the coactivation zone, thus making the mechanism of {*RC*; *k*} synergies available to stabilize the action. Further in this part, we will generalize the idea of {*RC*; *k*} synergies and the associated role of coactivation to actions by multijoint effectors.

Coactivation, from the mechanical point of view, is wasteful: It uses energy to activate opposing muscles that produce little change in resultant forces. So, the observed levels of muscle coactivation may reflect a trade-off between two factors, frugality in organizing actions and ensuring that {*RC*; *k*} synergies can be used to stabilize those actions. Different subpopulations solve this trade-off differently depending on the external conditions of task performance and person's perception of the task as being more or less challenging.

The switch from the muscle level to the joint-level control becomes less trivial if one considers the fact that most joints are spanned by multiple muscles. For example, the elbow joint with one axis of rotation, flexion–extension (let us ignore the pronation–supination action shared between the elbow and wrist joints), is spanned by three flexors and three extensors. Hence, the mapping between the control variables at the joint level, *r*-command and *c*-command per axis of rotation, and potentially many muscle-level control variables, $\{\lambda_i\}$, becomes nontrivial. Indeed, there is an infinite number of $\{\lambda_i\}$ combinations that can successfully produce a {*r*; *c*} pair at the joint level. This is an example of motor redundancy, which we consider as abundance. Are there multi-λ synergies stabilizing a desired $\{r; c\}$ pair? This seems feasible but, so far, there have been no experimental studies addressing this question directly primarily because of the lack of tool to measure individual λ values (lack of *lambda-meters*!).

Although the introduction of joint-level control variables, the $\{r; c\}$ pair, is logical, intuitive, and theoretically justified, there is little evidence for joint-level representation of movements. In fact, studies of muscle-level, joint-level, and movement-level representations in the primary motor cortex confirmed only the muscle-level and movement-level representations (Kakei et al., 1999, 2003). No evidence for joint-level representation was found in neuronal activation patterns. This result requires independent confirmation; as of now, it seems safe to consider the joint-level control as a useful theoretical construct.

5.3 Control at the effector level

The idea of control with spatial RCs can be generalized for movements of multijoint effectors up to whole-body movements in space (reviewed in Latash, 2010; Feldman, 2015). This is illustrated schematically in Fig. 5.11. The figure shows a hungry donkey attracted by a carrot. The carrot defines a RC in the external three-dimensional space. The donkey tries to ensure that the AC of its mouth coincides with the RC defined by the carrot. If the carrot cannot be reached by head motion, the donkey will have to solve the problem of moving the whole body to bring the head to the RC. When the donkey was a baby, it learned that trying to reach a distant piece of food without moving legs is unproductive: It leads to the body flopping on the ground. So, the carrot in

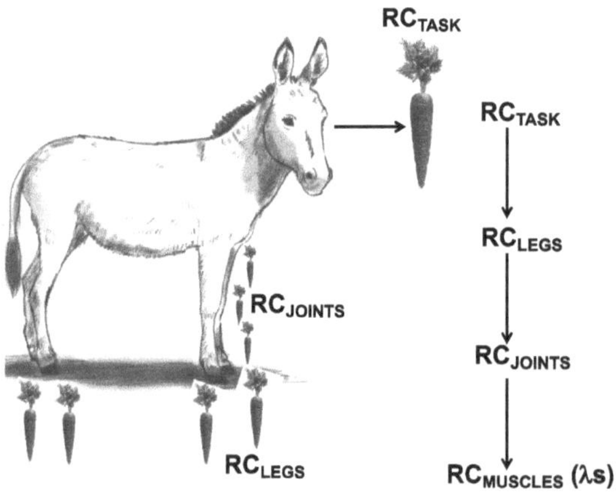

Figure 5.11 The carrot defines a referent coordinate (RC) in the external three-dimensional space for the hungry donkey. To reach the carrot, the donkey has to move the legs. So, smaller carrots define RCs for the legs, still smaller carrots—for the joints; for the muscles a carrot = λ, etc.

the external space has to be converted by the donkey's brain into a set of minicarrots (RCs) for the individual legs; this can be done with participation of a central pattern generator for locomotion in the spinal cord.

To move the legs, however, the donkey has to produce coordinated individual leg joint rotations. So, each minicarrot has to produce a set of microcarrots for the individual leg joints. As mentioned in the previous text, each joint is likely to be crossed by a large, abundant, number of muscles. So, each microcarrot should produce nanocarrots for individual muscles, which are equivalent to λs for those muscles. Of course, one can continue and state that each lambda has to be mapped on minilambdas for individual motor units, but let us stop the scheme in Fig. 5.11 at the level of individual muscles.

This scheme involves a hierarchy of control levels linked to each other via few-to-many (abundant) transformations. Theoretically, each of those transformations may be associated with synergies making sure that the combined action at the hierarchically lower level is close to the desired overall outcome prescribed at the higher level. Inother words, there may be cascades of synergies stabilizing ongoing action at different levels of the central nervous system. Of course, synergies are expected only in transformations when accuracy of the outcome in the unpredictable environment is important. Stability of corresponding salient performance variables is functionally crucial. For example, reaching the carrot with the head is definitely important for the hungry donkey. Is it important to stabilize trajectories of individual legs? They should be compatible with stability of the body, but actual hoof placement might not be that important. Are individual joint trajectories important? Maybe not as much although, during its life, the donkey has probably developed comfortable leg configurations (combinations of joint coordinates) and their trajectories preferred during walking.

Organizing synergies at each of the mentioned transformations is not impossible but may be nontrivial. An example of a trade-off between levels within a hierarchy has been considered earlier: When a person presses with two hands, two fingers per hand, to produce a desired level of total force produced by the four fingers, the two hands are organized into a synergy stabilizing total force, but the finger pairs in each hand show no synergies stabilizing that hand's contribution to total force (Gorniak et al., 2007). This is intuitively clear from the definition of *synergy*. Indeed, to qualify as a force-stabilizing synergy, the hand forces have to covary negatively across trials (as illustrated in Fig. 5.12A). But this means that variance within the UCM ("good variability") for total force is high, which means that each hand's intertrial variability is also comparably high. However, hand force variance is by definition "bad" when considered at the lower hierarchical level, i.e., at the level of two fingers generating the hand force. It is still possible to have two-finger synergies stabilizing that hand's force, but this may require very large magnitudes of variance along the corresponding UCM, possibly going beyond the available range of finger forces. In fact, studies of more functional hand tasks (holding an object in the air; Gorniak et al., 2009) have shown that some of the potentially important variables, such as load-resisting (vertical) force and resultant moment of force, are stabilized at two level of the hierarchy. In contrast, force applied orthogonal to the object's surface (grip force) is stabilized only at the higher level. In those studies, the upper level of the

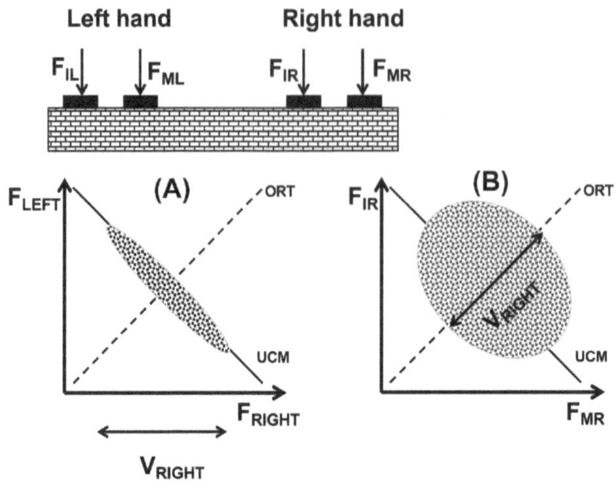

Figure 5.12 (A): When a person uses two fingers per hand to produce a desired level of total force, the two hands are organized into a synergy-stabilizing total force. This means that V_{UCM} in the space of hand forces (F_{LEFT} and F_{RIGHT}) is high, leading to large force variance for each hand (e.g., large V_{RIGHT}). (B): This naturally translates into large V_{ORT} at the individual hand (two-finger) level, in the space of the index (F_{IR}) and middle (F_{MR}) finger forces making it hard to organize a two-finger synergy stabilizing that hand's force.

hierarchy was defined as the one where task-related constraints are shared between the actions of the thumb and the *virtual finger*, an imagined digit with the mechanical action equal to the combined action of the four fingers (Arbib et al., 1985; MacKenzie and Iberall, 1994). At the lower level, action of the virtual finger is shared among the individual fingers.

Neurophysiological mechanisms of synergies are all but unknown. Likely, such mechanisms are distributed and act at different levels of the neural hierarchy. Earlier, a few examples of spinal mechanisms were discussed, which may be viewed as synergies stabilizing such variables as muscle activation level and the equilibrium state of a muscle acting against a constant external load. Whether there are similar mechanisms ensuring synergic mapping of $\{r; c\}$ joint-level variables on $\{\lambda_i\}$ muscle-level variables remains unknown. Given that such transformations are involved in natural limb and body movements, it is feasible that stability of the transformations is ensured by neural circuits, possibly at the spinal level, that may be shared by all limb and whole-body movements.

A general scheme of synergic transformations was suggested about 15 years ago (Latash et al., 2005). This scheme is built on feedback loops similar to those described for the system of Renshaw cells but with one important difference: The gains in the loops are assumed to be controllable and able to change sign. Fig. 5.13 illustrates such a scheme for a hypothetical system of four elements (E1,..., E4) contributing to a common output representing the sum of the four elemental variables. Each of the elements is controlled by a group of neurons (shown as single cells in the figure), which project also on small cells that exert effects on all the neurons of all four groups.

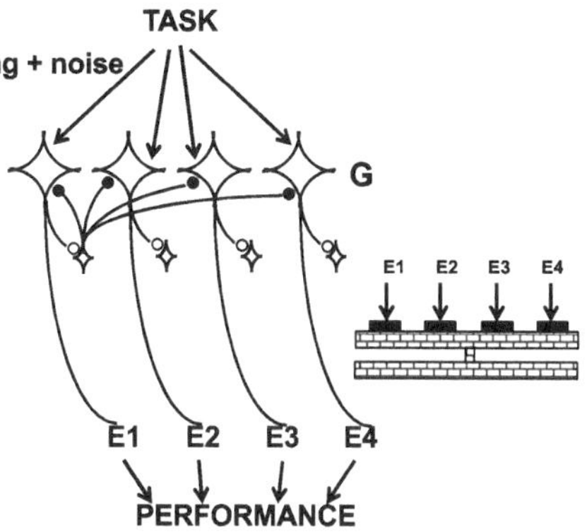

Figure 5.13 A hypothetical system of four elements (E1,..., E4) contributing to a common output. Each of the elements is controlled by a group of neurons (shown as single large neurons), which project also on small cells that exert effects on all the neurons of all four groups. The action of those small neurons can be described with a 4 × 4 gain matrix, **G**. Changing the magnitude and sign of the entries of the **G** matrix makes the system able to stabilize not only its total output but also other variables, for example, total moment of force produced by four fingers with respect to a pivot located in-between the points of finger force application (see the insert).

The action of those small neurons (similar to that of Renshaw cells, cf. Fig. 4.11) in the scheme in Fig. 5.13 can be described with a 4 × 4 gain matrix, **G**. If all the entries of the matrix are negative, the scheme is expected to produce covariation of the neuronal groups stabilizing their summed effect, e.g., as during steady-state total force production by four fingers pressing in parallel. Changing the magnitude of the entries of the **G** matrix can vary the strength of covariation. If some of the entries are allowed to change sign (become positive), the system becomes able to stabilize other variables, for example, total moment of force produced by the fingers with respect to a pivot located in-between the points of finger force application (see the insert in Fig. 5.13).

Feedback loops illustrated in Fig. 5.13 have been addressed as back-coupling mechanisms. They can be located within the central nervous system (as in Fig. 5.13) or involve feedback loops from sensory receptors producing information relevant to salient performance variables in specific tasks. A number of studies have demonstrated the importance of visual information for synergies in tasks where visual feedback was provided on task-specific performance variables (Ambike et al., 2016b; Parsa et al., 2016, 2017; Reschechtko and Latash, 2017). Removing this visual feedback led to major deterioration of performance involving unintentional drifts in the performance variable magnitude (see also Vaillancourt and Russell, 2002) associated with disappearance of respective synergies. On the other hand, in more natural tasks, strong performance-stabilizing synergies are seen without visual feedback on the respective variables. For example, imagine a person holding a vertically oriented object with the four fingers and the thumb acting in the so-called *prismatic grasp*: The contact with the object is limited to the digit tips, and the four fingers oppose the thumb. In this digit configuration, there are strong synergies stabilizing the resultant vertical (load-resisting) force, horizontal (gripping) force, and total moment of force (reviewed in Zatsiorsky and Latash, 2008;

Latash and Zatsiorsky, 2016). This is seen despite the fact that the subject receives no visual feedback on all those mechanical variables but simply tries to keep the object motionless. Of course, keeping the object motionless implies stability of the mentioned mechanical variables, but the example shows that an instruction combined with some information on performance (for example, coming from somatosensory receptors in the digit tips, the hand, and the arm) may be sufficient to form relevant performance-stabilizing synergies.

The last example suggests that the organization of *motor* synergies stabilizing performance may be tightly coupled with *sensory* information relevant to the task. This topic may be seen as a particular example of the general field of action–perception coupling. It suggests that, to produce stable actions, one has to possess stable percepts of relevant variables—a topic addressed in Part 6 of the book.

5.4 Drifts in performance as natural behaviors

One of the direct consequences of the idea of control with RCs is that muscle activations and active force production are consequences of the difference between RC for the effector and its AC. If RC = AC, the effector is expected to be in an equilibrium state with no active force production. Of course, if the effector continues to move, AC changes, and the equilibrium is lost resulting in the generation of muscle forces. For example, a pendulum shows oscillations because, when its AC = RC (this happens when the pendulum passes the vertical), movement continues and causes a deviation between AC and RC, leading to forces that slow down and then reverse the movement. Another way to disturb the equilibrium and to generate active muscle forces is to change RC. In either case, force magnitude changes with the difference (AC − RC), and these forces are directed to move the effector's AC toward the current RC. If the effector cannot move because of external constraints, e.g., as during pressing on a surface with a finger, AC cannot change, and the difference (AC − RC) is translated into active force production against the stop. In a linear approximation, the force may be viewed as proportional to (AC − RC) with a stiffness-like coefficient k. This brief summary ignores, for simplicity, the possibility of changing k with the earlier introduced coactivation command (c-command), which defines spatial ranges where opposing muscles are both active.

Fig. 5.14 illustrates this description for unidirectional action along a coordinate X. When RC = AC, the effector (shown in Fig. 5.14 as a ball) rests on the bottom of a potential well. A change of RC shifts the well, and the effector moves along the gradient toward the new location of the bottom of the well, i.e., it performs a movement. If motion is blocked (as in the bottom panel), active force is produced against the stop. Such movement and force production caused by RC shifts are expected to occur at a minimal time delay caused primarily by conduction delays along neural pathways and the time it takes for the muscle to generate force in response to a change in its activation level. Overall, these times are on the order of 0.1 s.

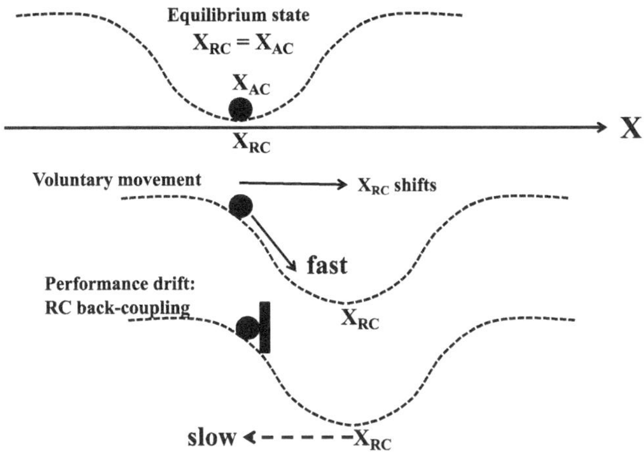

Figure 5.14 Imagine motion along a coordinate *X*. Top: When referent coordinate of the effector coincides with its actual coordinate, RC = AC, the effector (shown as a ball) is in equilibrium state on the bottom of the potential well. Middle: A change of RC shifts the well to a new spatial location, and the effector moves along the gradient toward the new coordinate of the bottom of the well. This is how voluntary movements are produced. Bottom: If motion is blocked, active force is produced against the stop, and RC shows an unintentional slow drift toward AC, leading to performance drift. This mechanism has been addressed as RC back-coupling.

5.4.1 Drifts in force production tasks

A number of observations have suggested that RC can shift unintentionally if the effector is prevented from moving to RC for a relatively long time, on the order of a few seconds (Wilhelm et al., 2013; Reschechtko et al., 2014; Zhou et al., 2014; 2015). For example, if a person is asked to press with a finger and produce a certain comfortable force level with the help of visual feedback on the force magnitude, removing visual feedback leads to a force drift toward lower magnitudes (Slifkin et al., 2000; Vaillancourt and Russell, 2002; Ambike et al., 2015). This happens even if the subject of such an experiment honestly tries to follow the instruction: "continue doing what you have been doing." Finger force can drop by a large magnitude, up to 30% of the initial value, and the subject is unaware of this force drop. Such force drifts have been approximated with exponential functions with the characteristic times ranging from 10 to 20 s. As such, this is a relatively slow process and its effects can be seen only during tasks with prolonged episodes of RC $\neq$ AC.

In typical experiments with force drop, if the subject is asked to relax for a couple of seconds and then reproduce the initial force based on memory or to produce this force with the contralateral effector, no drifts to lower forces are seen (Solnik et al., 2017; Reschechtko et al., 2018). This suggests that the subject has not forgotten the initial force in contrast to the original hypothesis that force drifts were consequences of problems with working memory (Vaillancourt and Russell, 2002). Note that the memory-based hypothesis has been indirectly supported by studies of brain electrical and metabolic activity (Vaillancourt et al., 2003; Coombes et al., 2011; Poon et al., 2012) and a study of patients with Parkinson's disease who are known to display working memory problems (Vaillancourt et al., 2001). However, the mentioned results with force reproduction suggest that the drift originates somewhere in-between the memory-based specification of the efferent neural process and actual muscle activations. Within the scheme in Fig. 5.14, it suggests an unintentional drift of RC toward AC and/or a drift of *k* toward lower magnitudes leading to a force drop.

The hypothetical slow drift of RC toward AC has been addressed as RC back-coupling. It may be seen as a reflection of the natural tendency of natural systems to move toward states with lower potential energy: If AC cannot move toward the bottom of the potential well, the bottom of the well starts drifting toward AC.

The magnitude of the force drop is larger for higher initial force levels, and it disappears (or is even reversed) when the initial force level drops to small values (under 5% of maximal voluntary contraction). This latter effect is possibly due to peripheral properties of pressure-sensitive receptors that can adapt to steady-state pressure and reduce their activation level (e.g., Iggo and Muir, 1969). Indeed, sensory endings in the fingertip, sensitive to maintained deformation, show adaptation, i.e., a drop in their activity with time if the external pressure remains constant. Such effects could be expected to lead to a force drift toward higher values: Smaller signals from pressure-sensitive receptors are likely to be interpreted by the subject's central nervous system as a force drop (which does not happen) and result in corrective actions leading to force increase. This factor is expected to dominate when the difference (RC − AC) is small and the hypothetical effects of RC back-coupling are also small, as in tasks with low magnitudes of force production.

Likely, there are more factors contributing to the actually observed unintentional changes in motor performance. Depending on specific conditions, some of the factors dominate and define characteristics of the drifts. For example, RC back-coupling probably happens at all times, but its effects are seen only over relatively large time intervals and only in the absence of performance corrections. As studies of finger force production show, the natural human perception of finger force is rather poor and unable to discern force deviations, even when these deviations reach large magnitudes, up to one-third of the initial force level. In contrast, perception of joint configurations and locations of effectors in space is rather accurate. So, even in the absence of visual feedback, no major drift in body configuration takes place, and one has to apply perturbations to observe effects of RC drifts (Zhou et al., 2014; 2015).

A few studies have shown that transient perturbation applied to the involved effector could provoke unintentional drifts in performance, which were not observed otherwise. In one of the experiments, discussed in more detail later, the subject held a hand position against a constant external force provided by a programmable robot and was instructed not to react to possible force changes. When a perturbation was applied by the robot, the hand moved away from the body. If the perturbation was immediately removed, the hand moved back and stopped close to the original coordinate. Even a relatively brief dwell time (1 s) before removing the perturbation led to a major violation of equifinality, i.e., error in the original hand position, which was interpreted as a consequence of an unintentional drift of the hand RC triggered by the perturbation-induced increase in the (AC − RC) difference.

During force production tasks, positional perturbations were applied by a specially constructed device called "*inverse piano*" (Martin et al., 2011). When a person plays piano, he or she pressed on the keys with fingers. In the inverse piano, keys press on the fingers and move them (the keys are powered with programmable linear actuators). Fig. 5.15 illustrates results of an experiment when the subject was asked to keep constant pressing force by a set of fingers and then not to react to possible finger motion. The inverse piano lifted the fingers smoothly, kept them at the new coordinate for 2 s, and then lowered them back to the original position. There was a major drop in the finger force during the dwell time reflected in the lower force level after the transient perturbation was over. This is also an illustration of violation of equifinality caused by a drift in RC.

Unintentional drifts in performance are not limited to steady-state tasks. In particular, they have been described for cyclical finger force–production tasks (Levy-Tzedek et al., 2011; Ambike et al., 2016c) and for cyclical whole-body sway tasks performed by standing persons (Rasouli et al., 2017). In particular, when a person is performing accurate cyclical changes in finger force with the help of visual feedback on the actual force and an auditory metronome, turning the visual feedback off leads to two consequences. First, the peak-to-peak force magnitude oscillations increase within 1–2 s and stabilize at a new magnitude. Somewhat surprisingly, the magnitude of the amplitude increase does not depend on the initial magnitude of the amplitude, suggesting that this process

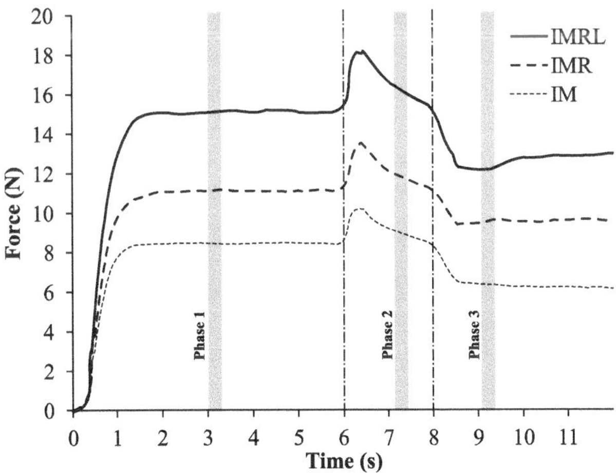

Figure 5.15 Drifts in force caused by a transient positional perturbation of the fingers with a 2-s dwell time. The task was performed by two (index—I and middle—M) fingers, three fingers IMR (R—ring), and all four fingers (IMRL).
Reproduced by permission from Reschechtko, S., Zatsiorsky, V.M., Latash, M.L., 2015. Task-specific stability of multi-finger steady-state action. Journal of Motor Behavior 47 365–377. Copyright by Taylor and Francis Publishers.

is not defined by the existence of a preferred amplitude of force oscillations (Ambike et al., 2018). At the same time, there is a much slower drift in the midpoint of the force cycle toward lower magnitudes similar to the drift seen during steady-state force production (Fig. 5.16). These observations suggested the existence of two kinds of drifts: A slow drift caused by processes within the UCM for the salient variable and a fast drift caused by processes within the orthogonal to the UCM (ORT) space (see later).

5.4.2 Complex force and phase drifts

More complex and counterintuitive patterns of performance drift were observed in two-finger cyclical force production tasks (Reschechtko et al., 2017a; Hasanbarani et al., 2018). In those studies, the subjects were asked to produce cyclical force changes while being paced by an auditory metronome at all times at a comfortable frequency. In one of the tasks, the subjects tried to produce cyclical changes in total force while maintaining the sharing of total force (relative contribution of each finger to total force) between the two fingers. In the other task, the instruction was to keep total force constant while producing cyclical changes in sharing. Note that the first task required finger force changes in-phase, and the second one required finger force changes out-of-phase. These two regimes are known as most stable across a variety of tasks (Schöner and Kelso, 1988; reviewed in Kelso, 1995).

Each trial started under full visual feedback on both force and sharing; subsequently, feedback on the variable that was instructed to stay constant was frozen (i.e., stopped producing salient information), and, after some time, feedback on the other variable was also removed. In both tasks, turning off visual feedback on total force elicited a drop in the midpoint of the force cycle and an increase in the peak-to-peak force amplitude as in the aforementioned studies with single-finger force production (Ambike et al., 2016c). The most surprising observation was that, without visual feedback, the two force time series started to deviate from the in-phase pattern and from the out-of-phase pattern during the first and second task, respectively (Fig. 5.17). In a way, subjects started to "canter" (cf. Robilliard et al., 2007) with the two fingers without realizing what they were doing. This was a very unexpected result, given the mentioned high stability of in-phase and out-of-phase actions and low stability of any other phase regimes.

Consistent phase shifts during bilateral human actions can be seen under asymmetrical loading of the moving effectors (Sternad et al., 1995, 1999; Russell et al., 2016). However, the conditions in the mentioned experiment were symmetrical and never modified. The fact that the subjects produced force without moving could play an important role because force production tasks are particularly sensitive to visual information (Lafe et al., 2016a,b). The drifts could be facilitated by the relatively poor perception of force based on signals from somatosensors: Indeed, even when humans have undistorted feedback from all the peripheral force-sensitive receptors in the arm and hand, they are unaware of the force change by up to 30% during the force drift observed typically after turning off the visual feedback on force (Vaillancourt and Russell, 2002: Ambike et al., 2015). Along similar lines, when humans cannot detect the shifts in relative phase between the outputs of two effectors because of inadequate

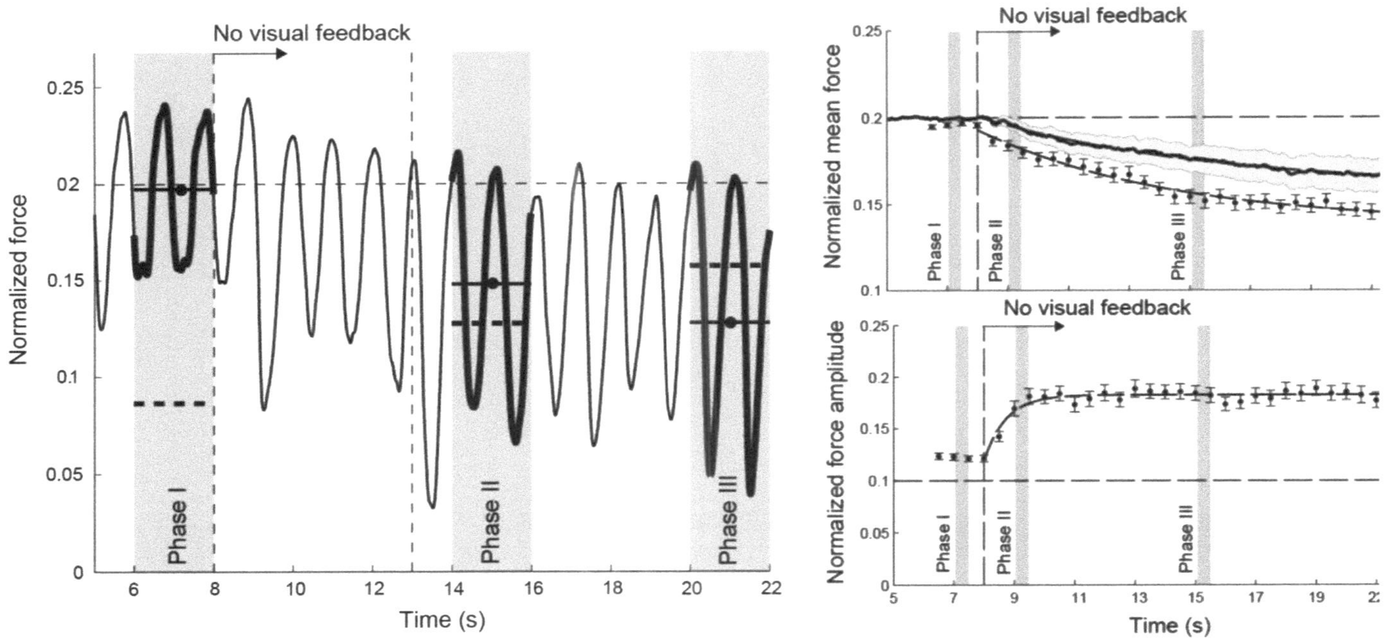

Figure 5.16 During accurate cyclical force production tasks, turning visual feedback off leads to drifts in both major characteristics of the force time profile: The midpoint of the force cycle drifts to lower values (similarly to the force drifts during steady force production), whereas the peak-to-peak amplitude increases. The left panel shows typical performance of a subject. The right panels show averaged across subject changes in the midpoint of the cycle (top) and peak-to-peak amplitude (bottom). Averaged force drift during the constant force production task is show in the top panel for comparison. Note that the cyclical task is performed with the auditory metronome on at all times.
Modified by permission from Ambike, S., Mattos, D., Zatsiorsky, V.M., Latash, M.L., 2016. The nature of constant and cyclic force production: unintentional force-drift characteristics. Experimental Brain Research 234 197–208. Copyright by Springer.

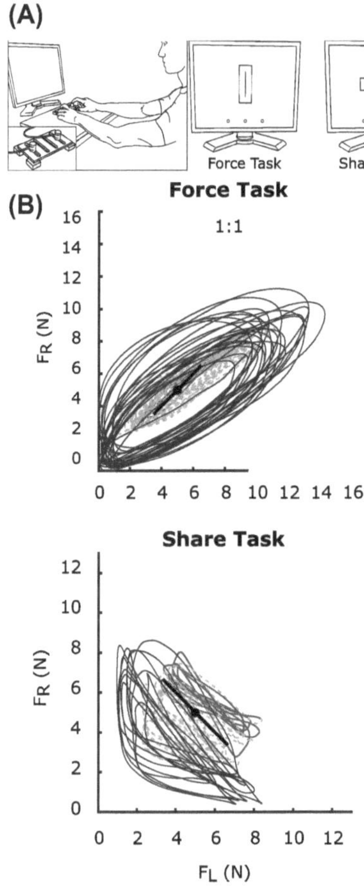

Figure 5.17 The subjects performed an accurate task that required cyclical changes in the total force (Force task) or sharing of the force between two fingers (Share task). (A): An illustration of the setup and the feedback during the Force task and Share task. (B): Performance by a typical subject. Y-axis shows right finger force (F_R), whereas X-axis shows left finger force (F_L). Progression of time is illustrated by line color and style: the first 8 s are shown with *solid blue lines* (gray lines in print version) (under full visual feedback); 8–23 s are shown with *dashed green lines* (light gray lines in print version) (feedback on the non-task variable was suppressed); 23–35 s are shown with *red lines* (dark gray lines in print version) (no feedback). The *black, thick line* represents ideal performance.
Modified by permission from Reschechtko S, Hasanbarani, F., Akulin, V.M., Latash, M.L., 2017. Unintentional force changes in cyclical tasks performed by an abundant system: Empirical observations and a dynamical model. Neuroscience 350 94–109. Copyright by Pergamon Press.

perceptual information, they may be unable to correct phase deviations when they occur (Wilson et al., 2003, 2010; Wilson and Bingham, 2008).

Within the physical approach to biological movements, performance drifts reflect the natural tendency of physical systems to move toward stable states as a result of drifts in their parameters. During cyclical force production, there are at least two parameters per effector (per finger) that can drift at different rates; these drifts are reflected in the relatively fast changes in the peak-to-peak amplitude and in the relatively slow changes in the midpoint of the force cyclical changes (as shown in Ambike et al., 2016c).

Such drifts may happen at different rates in the dominant and nondominant limbs, which would be compatible with the dynamical dominance hypothesis (Sainburg, 2002, 2005). This hypothesis assumes that the dominant hemisphere is specialized for the control of fast actions, whereas the nondominant hemisphere is specialized for stability of steady states. For example, during hand force production tasks, turning the visual feedback off causes smaller force drift magnitudes in the nondominant hand (Parsa et al., 2017a), which may be seen as a reflection of higher force stability. In contrast, preparing to a quick action (such as a quick force pulse) and performing it shows advantage of the dominant hand (Zhang et al., 2006; De Freitas et al., 2018). In the mentioned original study, a model with parameters drifting at different rates in the two hands was able to reproduce some of the main results including the phase drift.

The hypothesis that the phase drift is due to the hemispheric asymmetry was tested in a follow-up study with force production by two fingers of one hand only (Hasanbarani et al., 2018). Indeed, if the observed counterintuitive phase drifts were due to differences between the dominant and nondominant hands, two fingers within a hand are not expected to show such drifts. The results, however, did not confirm this prediction: Similar drifts were seen in the interfinger phase away from both in-phase and out-of-phase. At this moment, the surprising phase drift remains unexplained (BMC).

5.4.3 Drifts in kinematic tasks

Unintentional drifts in performance have been observed in a variety of tasks not involving accurate force production. During hand positioning tasks, perception of the hand position in space is rather accurate, and no drifts are seen in the absence of visual feedback. For example, if you hold an object in the extended hand, the hand RC is higher than its AC; this allows counteracting the effects of gravity. If RC is drifted toward AC, the hand would move down. This does not happen, however, suggesting that such drifts are prevented and/or their effects are quickly compensated for, such that the hand position remains undisturbed. These observations may reflect the functional importance of accurate hand movements without visual control in everyday life. However, in experiments with smooth, transient changes in external force, drifts in the hand position can be observed.

Transient perturbations are known to lead to equifinality, i.e., return of the effector to the original coordinate. Equifinality has been observed in many studies (Bizzi et al., 1976; Kelso and Holt, 1980; Schmidt and McGown, 1980; Latash and Gottlieb, 1990), although violations of equifinality have also been reported under certain special conditions such as being rotated in a centrifuge (Lackner and DiZio, 1994; DiZio and Lackner, 1995) or performing movements in an artificial environment with negative damping (Hinder and Milner, 2003). Note that the latter condition is never encountered in real life. The argument whether violations of equifinality are evidence against the EP hypothesis should be over by now (cf. Feldman and Latash, 2005). Indeed, the EP hypothesis predicts equifinality only under special condition such as no reaction of the person (or animal) to the transient perturbation and no changes in muscle properties. When such conditions are violated, no equifinality is expected; indeed, when drifts in control variables happen unintentionally, drifts in performance (violations of equifinality) should be seen.

Imagine a person who is asked to occupy a hand position in space against an external force vector (Fig. 5.18A). Then, the force is increased smoothly leading to a hand motion away from the body, followed by a short dwell time at the new force level, and then the force decreases back to the previous magnitude. The subject in this experiment is asked not to react to hand movements induced by the external force changes. These experiments showed two important results. First, when the transient perturbation had no dwell time at the new force level, the hand came back close to the initial position, i.e., it showed no major violations of equifinality. In contrast,

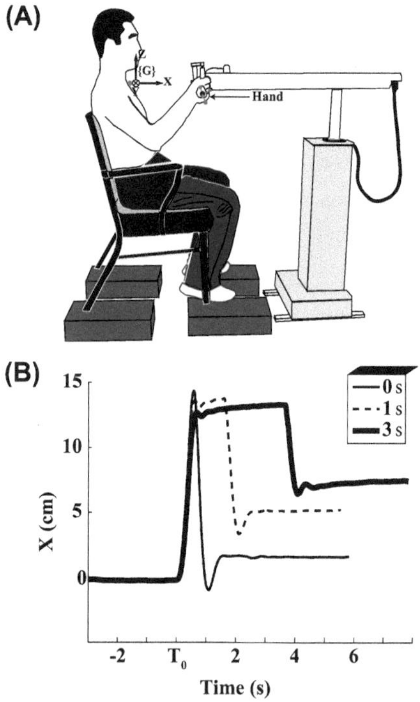

Figure 5.18 (A): A person is sitting and holding a handle against a constant force produced by a robotic hand. (B): The force shows a smooth increase, dwells for a short time, and then decreases back to the initial level. The final hand position differs from the initial one, and this difference increases with the dwell time.
Modified by permission from Zhou, T., Zhang, L., Latash, M.L., 2015. Characteristics of unintentional movements by a multi-joint effector. Journal of Motor Behavior 47 352–361. Copyright by Taylor and Francis Publishers.

individual joint positions showed large variance across trials aligned with the UCM for the hand location. In other words, the task-specific variable showed equifinality, whereas elemental variables did not (Zhou et al., 2014a). Second, when the external force dwelled at a new level before returning back to its initial level, the hand showed violations of equifinality: It moved toward the initial location but undershot it (Fig. 5.18B; Zhou et al., 2014b, 2015c). The magnitude of the undershoot was small for short dwell times and plateaued for the dwell times of about 1–2 s.

A follow-up study used additional hand perturbations to explore possible changes in its apparent stiffness and referent coordinate, RC and k (Zhou et al., 2015a, c). These studies have confirmed a drift in RC during the external force dwell time and a coupled drift in k, which prevented the hand from deviating from the new location during the dwell time when the external force was kept unchanged.

Another example of unintentional drift in performance was shown in a study of grip force changes with a smooth change in the size of the gripped object (Ambike et al., 2014). Typically, the grip force (the normal forces with zero resultant applied by the opposing digits to the gripped object) is somewhat higher than the minimal value to avoid slippage. For example, if a vertically oriented object is gripped by two digits (Fig. 5.19A), the normal force of each digit should be high enough to allow the application of vertical, load-resisting force: $\mu F_N \geq F_Z$, where F_N is normal force, F_Z is vertical force, and μ is the friction coefficient. The extra force above the slippage magnitude for F_N has been addressed as the *safety margin* (Johansson and Westing, 1984; Flanagan and Johansson, 2002; Jaric et al., 2006). When a subject gripped an object, and the object started expanding in the hand smoothly and then returned to its initial size (Fig. 5.19B), the normal forces of the opposing digits dropped by about 20% compared with their initial values so that the safety margin decreased to a dangerously low value.

Taken together, all the reviewed studies suggest that unintentional drifts in control variables, consequences of the RC back-coupling, are a universal phenomenon seen across tasks and effectors. Moreover, the results suggest the existence of two different time scales of such processes. Some of them are relatively quick with characteristic times on the order of 1–2 s. Other drifts are much slower and can be seen over 10–20 s time intervals. The difference in the characteristic times suggests that the drifts are likely to originate from processes happening in spaces with different stability properties. Indeed, in stable directions (those with high resistance to motion or high apparent stiffness), processes are expected to be relatively fast, whereas in less stable directions, processes are expected to be slower. These general considerations suggest a link between drifts at different speeds and processes in subspaces of different stability, namely the UCM and ORT (illustrated in Fig. 5.20 for a two-finger force production task).

The reviewed observations suggest that slow drifts in performance characteristics typically happen during steady-state tasks that involve no perturbations and no corrective actions by the actor. In contrast, fast drifts are observed when a steady-state action is perturbed or when a quick change in a performance variable is superimposed on its steady-state component (such as a quick oscillation about a nonzero midpoint). In fact, a typical slow drift in finger force speeds up rather dramatically if the finger is lifted smoothly and then lowered back to its initial position (as illustrated in Fig. 5.15;

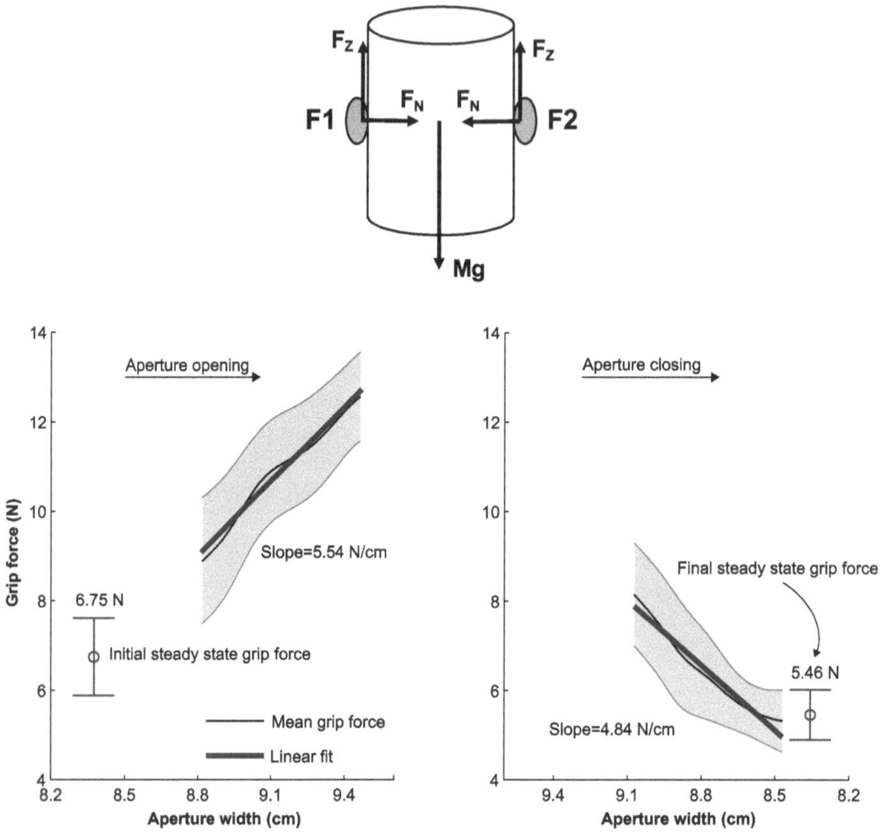

Figure 5.19 Top: When a vertically oriented object is gripped by two digits, the normal force of each digit (F_N) should be high enough to allow the application of vertical, load-resisting force (F_Z). Bottom: If the object starts expanding in the hand smoothly and then returns to its initial size, the normal forces of the opposing digits drop compared with their initial values. Modified by permission from Ambike, S., Paclet, F., Zatsiorsky, V.M., Latash, M.L., 2014. Factors affecting grip force: Anatomy, mechanics, and referent configurations. Experimental Brain Research 232 1219–1231. Copyright by Springer.

Wilhelm et al., 2014). Within this hypothesis, slow drifts originate in the UCM and are reflected in the ORT direction (they lead to a drift in the salient performance variable!) possibly because of a weak coupling between the two subspaces. Note that although UCM and ORT are mathematically orthogonal, processes within these spaces do not have to be independent. The hypothesized coupling between the UCM and ORT remains unexplored and, so far, BMC.

5.4.4 Drifts during standing

Some actions are strongly protected against unintentional drifts because of their crucial importance. These include keeping vertical posture in the field of gravity. To analyze

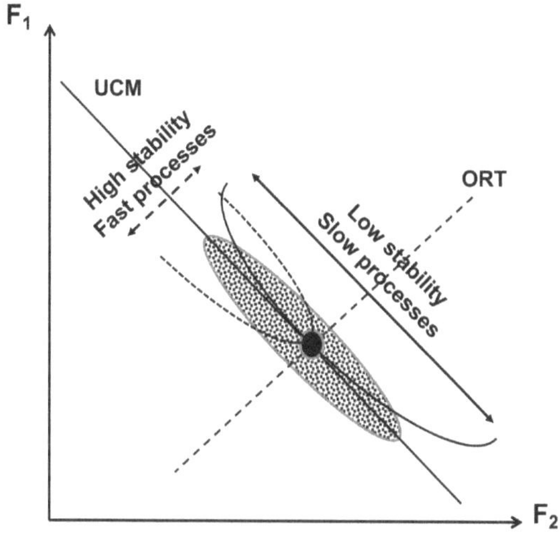

Figure 5.20 During a two-finger accurate force production task, the uncontrolled manifold (UCM) is shown with the *solid line* with negative slope, and the ORT—with the *dashed line* with positive slope. The high stability along the ORT and low stability along the UCM are shown schematically with potential fields with steep and shallow slopes, respectively. Lower stability along the UCM is expected to be associated with slower processes (such as drifts). Higher stability along the ORT is expected to be associated with faster processes.

the control of standing, we have to introduce a new notion that of referent orientation (RO). RO is a particular example of RC expressed in angular units. This notion has been used to address the neural control of rotational actions such as the production of accurate magnitudes of moment of force generated by the hand (Latash et al., 2010; Parsa et al., 2016, 2017).

Imagine now that a person is standing while leaning a little bit forward (Fig. 5.21). To keep balance, RO for the body has to be behind the actual person's trunk position. This allows generating active moment of force about the ankle joints counteracting the moment generated by the gravity force. According to the idea of RC back-coupling, RO is expected to drift toward actual body orientation (AO) leading to a decrease in (RO − AO) and a drop in the corresponding moment of force. The balance of moments should be lost leading to larger and larger lean forward until the person loses balance and falls down or makes a protective step. Such behaviors are not observed because of the strong posture correction mechanisms that try to bring the body back to its natural vertical orientation.

This theoretical competition of the two mentioned mechanisms has been studied in two types of studies. One of the studies explored unintentional drifts in posture during voluntary cyclical sway (Rasouli et al., 2017). In this study, the subjects were asked to sway, primarily about the ankle joints, between two visual targets. They were paced by an auditory metronome. Then, visual feedback was turned off and the subjects were asked to continue swaying. Similar to the mentioned studies of cyclical finger force production (Ambike et al., 2016c), the sway amplitude increased, while the midpoint could show drifts in different directions that were consistent within a subject but varied

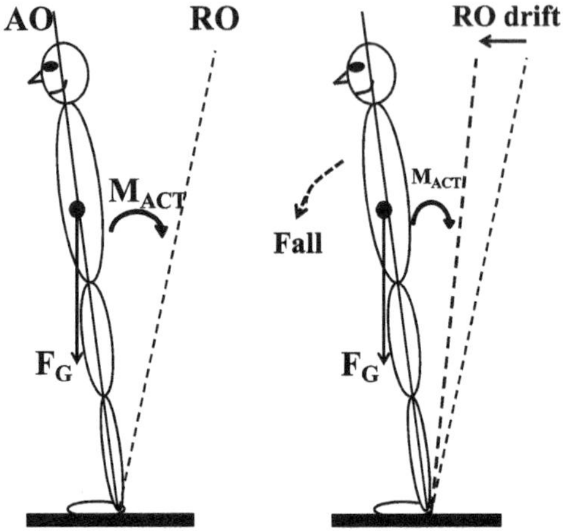

Figure 5.21 Imagine a standing person who is leaning a little bit forward. The control of vertical posture is associated with setting referent orientation (RO) of the body (RO is a specific example of referent coordinate in angular units). Left: When RO deviates from actual body orientation (AO), active moment of force (M_{ACT}) counteracts the moment produced by the gravity force (F_G). Right: If RO drifts toward AO, a decrease in (RO − AO) leads to a drop in M_{ACT}, and the body is expected to fall or the person makes a protective step.
Reproduced by permission from Yamagata, M., Popow, M., Latash, M.L., 2019. Beyond rambling and trembling: effects of visual feedback on slow postural drift. Experimental Brain Research 237 865–871. Copyright by Springer.

across subjects. Fig. 5.22 illustrates drifts in the sway midpoint for pairs of subjects who showed opposite drift directions that were consistent, however, across trials. Such behaviors were observed under different conditions, with the sway about the initial midpoint anterior (top panels) and posterior (bottom panels) to the comfortable standing coordinate. These observations suggest that individual subjects differed in the relative efficacy of the two mechanisms. One of them—the RC back-coupling—tended to move the midpoint of the sway away from the coordinate corresponding to natural vertical standing, whereas the other—the tendency to move toward the most stable posture—produced drifts of the midpoint toward the preferred vertical posture.

Another study explored characteristics of spontaneous postural sway. There are two polar attitudes toward the nature of spontaneous changes in a variety of mechanical variables seen when a person is asked to stand quietly. The first considers sway as a sign of imperfection of the postural control system, a consequence of some kind of intrinsic noise that has to be corrected using a variety of feedback mechanisms (Kiemel et al., 2002; Peterka, 2002; Maurer and Peterka, 2005). One of the first studies within this line of thinking viewed sway as body oscillation about the ankle joints produced by the stiffness-like properties of muscle crossing these joints (Winter et al., 1996). This hypothesis has not been supported in a study measuring the apparent stiffness of the ankle joint that showed values much lower than those needed to balance the body (Jacono et al., 2004). In addition, measurement of spontaneous joint deviations

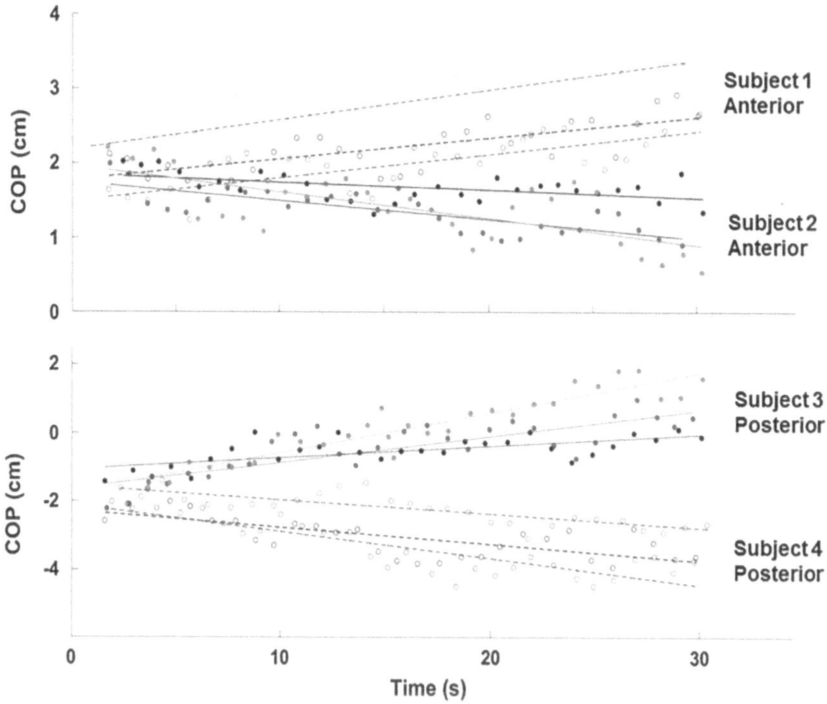

Figure 5.22 An illustration of the drift in the midpoints (M_{COP}) in the cyclical sway task for four subjects. Top panel: the task was performed about an anterior center of pressure (COP) position; bottom panel: the task was performed about a posterior COP position. M_{COP} for individual subjects are shown with *open and filled dots*; linear regression lines are drawn. Note the consistent drifts away from the natural standing COP coordinate (zero) in Subject 1 and Subject 4 and consistent drifts toward zero in Subject 2 and Subject 3.
Reproduced by permission from Rasouli, O., Solnik, S., Furmanek, M.P., Piscitelli, D., Falaki, A., Latash, M.L., 2017. Unintentional drifts during quiet stance and voluntary body sway. Experimental Brain Research 235 2301–2316. Copyright by Springer.

along the body vertical axis has shown large deviations in joints other than the ankle joints (Hsu et al., 2007). These joint deviations were organized into a synergy stabilizing coordinate of the center of mass, which would be of course impossible if only one joint motion contributed to the body sway (in that case, all variance in the joint space would be, by definition, V_{ORT}!).

The alternative view is that postural sway is reflection of a purposeful neural mechanism, possibly related to testing limits of stability of the vertical posture (Riccio, 1993; Mochizuki et al., 2006). This view is readily compatible with the idea of the neural control of posture with time changes in the body RO. Indeed, it predicts that sway should represent a superposition of two processes: One of them reflects purposeful changes in RO, whereas the other depends on peripheral mechanics defined by muscle–tendon properties and the action of reflex feedback loops. A method to explore these two sway components was developed by Zatsiorsky and Duarte

(1999, 2000) under the name of *rambling–trembling decomposition*. Rambling represents a trajectory of the center of pressure (COP, the point of application of the resultant vertical force acting on the body), computed as an interpolation of instantaneous EPs, i.e., the points when the resultant force in the selected direction, anterior–posterior or medio-lateral, is zero. Trembling is the difference between the COP trajectory and rambling.

Rambling, which is typically slower (with frequencies within the 0.1–0.4 Hz), has been viewed as predominantly a consequence of supraspinal control, whereas trembling (which is faster, 0.2–1.5 Hz) has been considered a reflection of peripheral mechanics and action of spinal reflex loops. The method of rambling–trembling decomposition has been used broadly and able to show differential effects on these two sway components of such factors as visual feedback (Danna-Dos-Santos et al., 2008), aging (Sarabon et al., 2013; Degani et al., 2017), typical and atypical development (Monteiro Ferronato and Barela, 2011; Speedtsberg et al., 2017), neurological disorders (Shin et al., 2011; Shin and Sosnoff, 2017), postural instability (Mochizuki et al., 2006), bed rest (Sarabon and Rosker, 2013), ballet dancing (Michalska et al., 2018), and even cannabis use (Bolbecker et al., 2018).

The idea of RC back-coupling allows predicting a third component, considerably slower than typical rambling: Note that characteristic times of unintentional drifts in earlier studies of finger force production were on the order of 10–20 s corresponding to frequencies under 0.1 Hz. A recent study of postural sway used two manipulations to reduce potential COP-stabilizing effects of the natural visual cues and, as a result, to facilitate postural drift (Yamagata et al., 2019b). In the first, the subjects stood quietly while matching a visual target on the monitor with a signal reflecting an irrelevant variable, the level of activation of a finger flexor muscle. In the second, the subjects stood with eyes closed. Closing eyes led to an overall increase in body sway without any special effect on its very low-frequency components, while using vision to perform an irrelevant task led to the emergence of visible, large-amplitude sway oscillations (illustrated in Fig. 5.23).

The very slow postural sway component was likely a consequence of an unintentional, comparably slow, RO drift toward the actual body orientation, a specific example of RC back-coupling. When the COP coordinate reached a threshold magnitude detected by the subject with the help of available sensory signals, somatosensory and/or vestibular, a postural correction was introduced. The drift continued until reaching the threshold magnitude again, followed by another postural correction, and so on (as can be seen in Fig. 5.23). Looking at external objects, with many vertical and horizontal cues, likely helped the subjects of that study to avoid the slow COP drift and/or correct it quickly and effectively.

5.4.5 Performance drifts in more natural actions

There are other examples of unintentional drifts in motor performance over very much different time intervals. For example, if a person is asked to walk toward an obstacle, step over it, and continue walking, there is a particular clearance between the foot

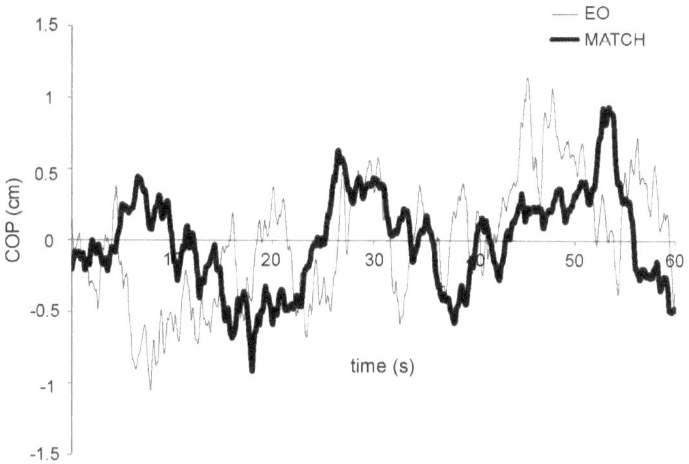

Figure 5.23 Examples of the center of pressure (COP) time series for a typical subject performing the task of quiet standing with eyes open (*thin line*, EO) and while matching the EMG of the finger flexors to a target (*thick line*, MATCH). Note the low-frequency cycles obvious in MATCH, not in EO, which are likely caused by the RC back-coupling and following postural corrections.
Reproduced by permission from Yamagata, M., Popow, M., Latash, M.L., 2019. Beyond rambling and trembling: effects of visual feedback on slow postural drift. Experimental Brain Research 237 865–871. Copyright by Springer.

stepping over the obstacle and the upper edge of the obstacle. If now this person is asked to perform this task numerous times in a row, the clearance tends to decrease and sometimes the person actually hits the obstacle with the foot of the swing leg (Heijnen et al., 2012, 2014). This is a rather complex task and, as of now, there are no explanations for this phenomenon that would go beyond operating with rather nonspecific or poorly justified notions such as boredom and minimization of energy.

Another example of an unintentional change in performance has been described under the name of "*slacking*" in studies of robotic-assisted rehabilitation after stroke (Reinkensmeyer et al., 2009; Smith et al., 2018). In those studies, robotic devices were used to help the patient perform movements that required involvement of paretic (weakened) muscles. Slacking was reflected in a progressive drop in the active contribution of the patients to movements: They preferred to allow the robot to do most of the job in performing actions, which was of course detrimental to the process of motor rehabilitation.

To summarize, there are many reasons why some actions readily show unintentional drifts in performance. An important lesson is that natural behavior of a physical system is expected to lead to unintentional drifts, which may be corrected if relevant sources of information and means of correction are available. It is possible that some movement impairments are associated with exaggerated RC drifts, as suggested, for example, by observations of larger force drifts in patients with Parkinson's disease (Vaillancourt et al., 2001; Jo et al., 2016).

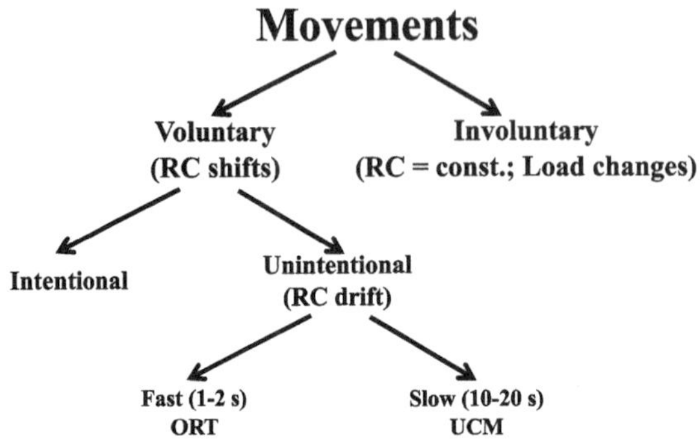

Figure 5.24 Classification of movements. Voluntary movements are associated with changes in referent coordinate (RC); involuntary movements are due to changes in external forces without a change in RC. Voluntary unintentional movements are produced by RC drifts that can be slow (likely originating in the UCM for the salient performance variable) or fast (likely originating in the ORT for that variable).

One can introduce the following schematic representations of different kinds of movements within the idea of control with RCs (Fig. 5.24). Voluntary movements are consequences of RC shifts with or without changes in the external loading conditions. If RC = constant and the external force changes, an involuntary movement takes place. For example, if you stand with a heavy book on the palm of a hand, and someone suddenly removes the book, an involuntary arm movement would take place to a new hand position in space.

Voluntary movements are associated with RC shifts and can be intentional and unintentional. These two classes differ in the origins of the RC change: A purposeful movement is produced by an intentional RC shift, while an unintentional movement (drift) occurs when RC changes without the person's knowledge. And finally, as described earlier, unintentional voluntary movements can be fast and slow, possibly related to the different speeds of processes within the UCM and ORT (note that all natural movements involve abundant sets of elements, e.g., muscles, such that UCM and ORT can always be defined).

5.5 Brain mechanisms of control with referent coordinates

The general idea of control with RCs suggests that numerous combinations of signals from the brain may be associated with a RC shift for effectors involved in an action. This idea received support in an experiment on animal preparations addressed as *decerebrate animals*. In that study (Feldman and Orlovsky, 1972), a surgery was performed on cats separating the brain stem from the more rostral parts of the brain including the large hemispheres; as a result, only signals from subcortical structures could reach the spinal cord and produce movements. The distal tendons of the target muscles in a hindlimb were separated from their natural points of attachment and used to apply external forces and record muscle length and force. The authors applied electrical stimulation to a number of brain structures and recorded changes in the force–length muscle characteristics. They reported changes in the threshold of the stretch reflex across all the stimulation conditions and did not observe independent changes in the slope of the force–length characteristics that would make such characteristics intersect. These results were interpreted as evidence for a universal rule: The descending control of muscles uses only one parameter, λ, which is equivalent to RC.

In a later study, also on cats, another group of authors (Nichols and Steeves, 1986) reported that stimulation of the magnocellular red nucleus could produce changes in the ankle joint mechanics suggesting intersection of stretch reflex characteristics of the muscles crossing the joint. Note that the red nucleus is the origin of a major descending tract, the rubrospinal tract, which is one of the main tracts delivering changes in activity of the cerebellum to the spinal cord. Although these results suggest that descending signals can change not only λ but also the shape of the force–length characteristic, potentially leading to their intersection, they remain compatible with the idea of control with RCs. The idea of intersecting RCs was later developed to create a model of movements in patients with dystonia (Latash and Gutman, 1994), a condition characterized by irregular twisting movements.

More recently, the cortical control of movements has been studied using healthy human subjects. Those experiments used the method of transcranial magnetic stimulation (TMS) to induce currents within the large hemispheres (reviewed in Rossini et al., 1994). This method is based on two physical phenomena that link electrical and magnetic effects (Fig. 5.25). The first one is that a change in the direction of electric current induces a magnetic field, which can penetrate biological tissues without causing any detectable effects. The second is that a change in the magnitude of magnetic field induces electrical currents in nearby structures able to conduct current. Based on these phenomena, the method uses a circular (or "figure eight") coil isolated from the body. A strong current in the coil induces magnetic field that can penetrate the skin, skull, and meninges. A quick change in the current magnitude produces a quick change in the magnetic field magnitude, which induces currents in conducting tissues in a vicinity of the coil. The currents stimulate axons running parallel to the cortical surface and making synapses on cortical neurons, ultimately leading to activation of the output neurons of the cortex (e.g., the pyramidal neurons in the primary motor cortex). Positioning and orienting the coil appropriately allows targeting specific subgroups of cortical neurons and producing activation in specific muscle groups, particularly in groups with large cortical representations such as hand and forearm muscles.

Muscle response to TMS depends on a number of factors such as excitability of the targeted cortical neurons and excitability of the alpha-motoneuronal pools. During voluntary muscle activation in isometric conditions (without motion), TMS-induced responses typically increase with the level of baseline muscle activation (Lim and Yiannikas, 1992; Ravnborg et al., 1991). This is an expected result given that excitability of a neuronal pool increases with the baseline pool activation, at least within a range of modest levels of activation (Danion et al., 2003). Note that according to

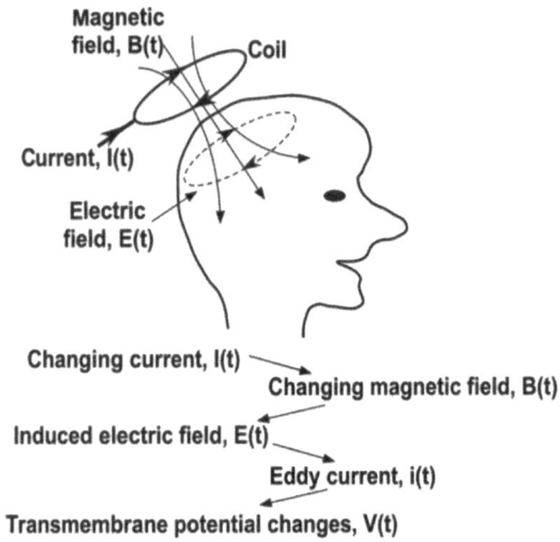

Figure 5.25 A scheme of brain stimulation using transcranial magnetic stimulator. The stimulator produces a pulse of current in the coil, **I**(t). The current changes direction in the coil and produces quickly changing magnetic field, **B**(t). The magnetic field penetrates the head and produces electrical field **E**(t), leading to currents **i**(t) that stimulate neural tissues and lead to action potentials. Ultimately, after synaptic transmission, these action potentials lead to a burst of activity in the corticospinal tract resulting in muscle contractions.
Reproduced by permission from Latash, M.L., 2012. Fundamentals of Motor Control. Academic Press New York, NY. Copyright by Elsevier Inc.

the idea of control with shifts in λ, muscle activation level reflects deviation of muscle length from λ. If cortical neurons encode values of λ, rather than muscle activation levels or peripheral mechanical variables, such as movement kinematics and/or kinetics (a traditional view: Evarts, 1968; Humphrey et al., 1970, 1978; Fetz and Cheney, 1980; Georgopoulos et al., 1992; Sergio and Kalaska, 1997; Hepp-Reymond et al., 1999), excitability of a cortical neuronal pool is expected to be different between conditions when the same muscle activation is observed at different muscle length values (i.e., at different joint positions). This prediction was tested by the group of Feldman (Raptis et al., 2010; Ilmane et al., 2013) in experiments with TMS applied at different wrist joint positions with matched muscle activation levels. As muscle activation levels were matched, excitability of the corresponding alpha-motoneuronal pools was also expected to be similar. Nevertheless, the TMS-induced responses differed significantly, and these differences could be interpreted as consequences of differences in the excitability of corresponding cortical neuronal pools. Hence, the authors concluded that cortical neurons encoded not muscle activation levels but magnitudes of λ.

In a recent study, the group of Feldman continued to explore the idea that descending signals from brain structures encoded magnitudes of RCs at the whole-body level: They studied postural adjustments in standing human subjects in response to stimulation of the vestibular apparatus (Mullick et al., 2018; see also Raptis et al., 2007). Those studies produced results suggesting that descending signals from the vestibular apparatus encoded RCs for the body orientation in space (i.e., RO). Taken together, all the mentioned observations are compatible with the basic hypothesis that signals from the brain to the spinal cord encode changes in RCs for the effectors that can range from individual muscles to the whole body.

5.5.1 Neuronal populations and encoding of referent coordinates

Earlier, we reviewed briefly a series of classical studies by Georgopoulos and his colleagues on possible encoding of performance variables, such as movement direction, in population vectors computed based on neuronal activity in the primary motor cortex (Georgopoulos et al., 1982, 1983; Kalaska et al., 1989, Section 4.4). Further studies on neuronal population vectors showed their correlation with other performance variables including force direction in isometric conditions, movement extent, levels of muscle activation, as well as more complex variables reflecting interjoint coordination (Kalaska et al., 1989; Lemon and Mantel, 1989; Maier et al., 1983; Sergio and Kalaska, 1998; Poliakov and Schieber, 1999; Cabel et al., 2001; Holdefer and Miller, 2002; Sergio et al., 2005).

Directional sensitivity is common across structures within the central nervous system beyond the primary motor cortex. It has been documented for neurons in other brain areas, such as the basal ganglia and the cerebellum (Crutcher and DeLong, 1984; Fortier et al., 1989; Pompeiano et al., 1997), as well as for neurons in the spinal cord, including alpha-motoneurons (Brink et al., 1984). Moreover, it has been suggested that any neuronal population is expected to show the basic feature: Correlation between the direction of a performance variable and the direction of the neuronal population vector (Mussa-Ivaldi, 1988), as long as its neurons possess two properties: (1) Individual firing frequency of neurons changes as a cosine function of the direction of a performance variable, and (2) the preferred directions of individual neurons cover the whole range of possible directions of that performance variable.

Note that a correlation between a particular performance variable and a variable measured at the level of neuronal activity does not mean that the latter caused the former. Moreover, as mentioned many years ago by Bernstein in a book written in the mid-1930s and published about 70 years later (Bernstein, 2003), neuronal activity in the brain cannot, in principle, encode movement mechanics. In a paper written about the same time (Bernstein, 1935), Bernstein suggested that neuronal activity could only encode topological features of movements but not their metrics with the help of so-called *engrams*. This insight was lost, unfortunately, for nearly half a century, and in later studies, a concept of *generalized motor program* (a development of the idea of engrams) was linked to such peripheral variables as muscle activations and forces (Schmidt, 1975, 1980).

Within the physical approach to biological movement, neuronal populations in the brain can encode time profiles of parameters of specific laws of nature that lead to observable patterns of performance variables only indirectly, with effects from the external force field, sensory signals, and other parts of the body. This general concept coupled with the idea of control with spatial RCs for the effectors suggests that neuronal activity in such brain areas as the primary motor cortex can be related to setting time profiles of corresponding RCs. In line with the mentioned Bernstein's insight, a recent series of studies by the group of Feldman have shown that corticospinal influences on muscle activations can be decoupled from peripheral variables such as baseline levels of muscle activation and muscle forces (Raptis et al., 2010; Sangani et al., 2011; Ilmane et al., 2013; Turpin et al., 2016). In other words, a standard stimulus applied to the primary cortex leads to quantitatively, and sometimes even qualitatively, different peripheral consequences depending on the baseline state of the peripheral motor apparatus.

These findings may not come as a complete surprise for those familiar with the phenomenon of *reflex reversal* (reviewed in Latash, 1993). Indeed, even a standard peripheral stimulus does not necessarily lead to a standard, localized reflex response. Depending on the initial posture of the animal and some other factors, the response can be seen in antagonist muscles or even in remote muscles crossing other joints (von Uexkull 1904; Magnus, 1924; Lisin et al., 1973; Forssberg et al., 1975; Latash and Gurfinkel, 1976; Duysens et al., 1990). So, even reflex loops, which do not extend beyond the spinal cord, do not prescribe emerging peripheral patterns given the actual current state of the animal.

Recently, Feldman (2019) has suggested that neuronal population characteristics in the primary motor cortex encode planned changes in the RC for the involved effectors, including the direction of RC shifts. In reproducible experimental conditions, these hypothetical RC(t) shifts lead to consistent changes in a variety of peripheral variables such as effector force vector, displacement, muscle activation, etc. A change in the experimental condition, however, is expected to lead to a change in the performance variable correlating most closely with the neuronal population vector. This is a logical hypothesis within the physical approach to motor control, which is compatible with a body of published data. The hypothesis is still waiting for detailed experimental verification.

5.6 Movement stability in the space of control variables

Experimental studies of control variables, such as RCs, have faced a number of problems. First, these variables are not directly measurable (as mentioned, there are no lambda-meters!) and have to be computed based on measurable mechanical and electrophysiological variables. Second, such computations necessarily involve approximate models of the relations between RCs and the measured variables. Third, and maybe most important, biological systems are active, which means that they react to external perturbations, which are commonly used during measurements, with changes in RCs. Such reactions are large and obvious for quick and large-magnitude external perturbations. This makes it impossible to use typical system-identification techniques that are commonly applied in engineering to define parameters in the assumed models: Parameters of the model and even its structure (the functional form of the corresponding equations) can change as a result of the system's reaction to a perturbation.

Indeed, imagine that you want to identify a parameter of a biological subsystem, e.g., an RC for an effector. If you apply a strong and quick perturbation, there will be reflex and reflex-like reactions (sometimes addressed as preprogrammed reactions, reviewed in Latash and Zatsiorsky, 2016) leading to changes in that parameter. So, states of the system before, during, and after the perturbation differ and cannot be easily compared to quantify the parameter of interest. If one applies a very quick perturbation and measures the states of the system within a very short time, shorter than the time delays involved in system's reactions, effectively one studies a deafferented system, i.e., a system with no effects of peripheral sensory information. Parameters of such a system may show large differences from those characterizing the intact system (cf. Nichols and Houk, 1976).

Therefore, one has to be sneaky and use not quick perturbations but smooth and relatively small changes in external forces. This method does not trigger reflexes and reflex-like reactions and allows observing natural behavior of the system with presumably unchanged parameters. The assumption of unchanged parameters in such studies so far remains untestable, although there were attempts to address it (Latash, 1994).

The general logic of such studies is as follows. You ask a subject to perform a simple motor task and instruct him or her not to correct possible errors in performance if they happen. Then, in some trials, the external force field changes smoothly leading to changes in movement trajectory and causing an error in performance. You assume that the time profiles of the control variables (RCs) are unchanged in such cases. Then, comparing performance variables across different perturbed and unperturbed trials allows reconstructing RC(t) time profiles (or their values if the task is a steady-state one). Of course, the process of reconstruction requires accepting a model that links performance variables measured in such an experiment to RC values.

This method was applied nearly 30 years ago to studies of single-joint movements in healthy humans and led to somewhat controversial results (Latash and Gottlieb, 1991, 1992). When the subjects performed unidirectional quick elbow movements, the method showed nonmonotonic changes in RC for the joint (illustrated in Fig. 5.26). These findings were criticized later (Gribble et al., 1998) as potentially reflecting the simplified, linear second-order model accepted in the original studies. In particular, the original study was criticized for assuming low magnitudes of damping and ignoring contributions to damping of reflex-mediated loops originating from velocity-sensitive peripheral receptors. This made the system underdamped (compatible with the published data, Cannon and Zahalak, 1982; MacKay et al., 1986; Lacquaniti et al., 1982; Bennett 1993) and resulted in the finding of nonmonotonic RC control. Accepting different (much higher) magnitudes of the damping was able to account for the observed behavioral patterns with monotonic RC changes.

Nevertheless, follow-up studies seemed to provide additional evidence compatible with the nonmonotonic RCs. The same method (and the same model) was applied to studies of cyclical movements and led to nonmonotonic changes in the peak-to-peak amplitude of RC(t) with a monotonic increase in the movement frequency (Fig. 5.27; Latash, 1992). The amplitude of RC(t) changes was relatively large, comparable with the movement amplitude, and during slow movements, it became nearly zero at an intermediate frequency and increased again for even faster

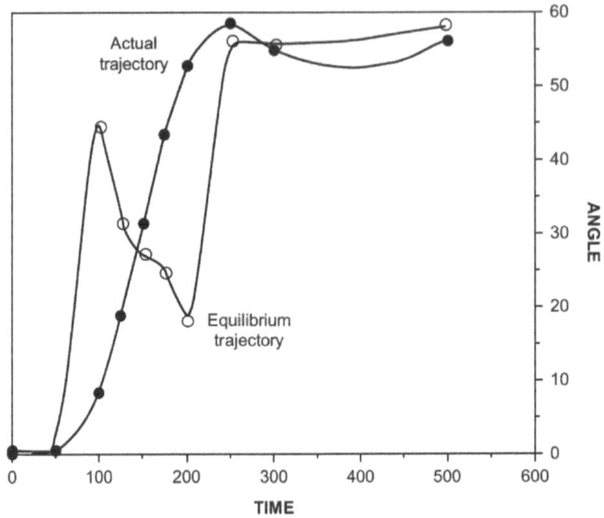

Figure 5.26 A healthy person performs a series of quick elbow flexion movements against a constant external torque. In some trials, the external torque changes smoothly leading to deviations of joint trajectories. Analysis across trials allows reconstructing equilibrium trajectory of the joint under the assumption that the subject did not correct movements. Modified by permission from Latash, M.L., Gottlieb, G.L., 1991. Reconstruction of elbow joint compliant characteristics during fast and slow voluntary movements. Neuroscience 43 697–712. Copyright by Pergamon Press.

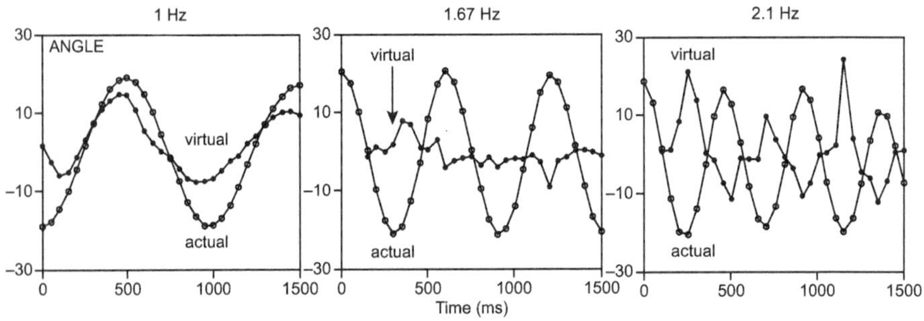

Figure 5.27 A healthy person performs cyclical elbow flexion movements against a constant torque. The method of reconstructing equilibrium trajectories (shown in the figure as "virtual trajectory") showed relatively large peak-to-peak changes in RC during slow movements, nearly flat RC(t) for an intermediate frequency, and an increase in the RC amplitude of changes for even faster movements.
Modified by permission from Latash, M.L., 1992. Virtual trajectories, joint stiffness, and changes in natural frequency during single-joint oscillatory movements. Neuroscience 49 209–220. Copyright by Pergamon Press.

movements. The frequency when RC(t) showed irregular changes with small peak-to-peak amplitude was estimated to be close to the natural frequency of the effector, which makes sense: Indeed, a physical object needs minimal ongoing control to move at its natural frequency, only to compensate for energy dissipation. Although the issue of monotonic versus nonmonotonic RC(t) trajectories remains unresolved, the method itself seemed to work reliably and led to consistent findings across subjects and task conditions.

5.6.1 Referent coordinate–based synergies

Recently, the idea of applying relatively slow, smooth perturbations has been developed to explore stability of actions in spaces of hypothetical control variables (Ambike et al., 2016a; b, 2018). Consider the control of a single effector, which is required to produce an accurate magnitude of force in isometric conditions, e.g., a finger pressing against a stop (Fig. 5.28A). The fingertip coordinate, X_0, is fixed, for example, at the

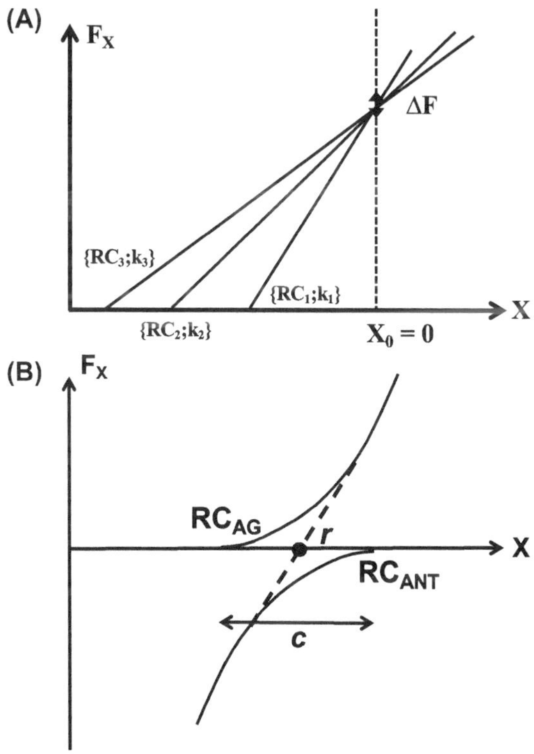

Figure 5.28 A: There are numerous combinations of referent coordinate (RC) and apparent stiffness (*k*) that allow producing the desired force magnitude within a target window (ΔF) in isometric conditions (effector coordinate $X_0 = 0$). B: The control of an effector can be described with setting magnitudes of RC for agonist and antagonist, RC_{AG} and RC_{ANT}, or, equivalently, as setting RC (by a change in the *r*-command) and *k* (by a change in the *c*-command) for the effector.

origin of the reference system, $X_0 = 0$, and the task has a specific error margin, ΔF. Any effector is controlled by at least two groups of muscles with opposing actions, agonist and antagonist. So, at the control level, there are two variables, RC_{AG} and RC_{ANT} (Fig. 5.28B). Each muscle group is characterized by a nonlinear force-coordinate characteristic, and the algebraic sum of the two characteristics is the effector characteristic (the dashed line in Fig. 5.28B).

When RC_{AG} and RC_{ANT} are shifted in the same direction by the same magnitude, the effector characteristic shifts parallel to itself along the coordinate axis. This type of control has been referred to as the reciprocal command (*r*-command, Feldman, 1980). If RC_{AG} and RC_{ANT} are shifted in opposite directions by the same magnitude, the effector characteristic rotates without a change in its intercept with the coordinate axis. This type of control has been referred to as the coactivation command (*c*-command, Feldman, 1986). So, any control of an effector can be described with setting magnitudes of RC_{AG} and RC_{ANT} or, equivalently, as setting magnitudes of the *r*- and *c*-commands. Note that the *r*-command shifts the referent coordinate for the effector, RC, whereas the *c*-command changes its apparent stiffness (k). As illustrated in Fig. 5.28A, there are numerous combinations of RC and k that allow producing the desired force magnitude. We arrive at a conclusion that, at the level of control, a single-effector force production task is abundant! This makes it possible to ensure stability of performance at the control level even if at the level of mechanics the task is not abundant. Does the central nervous system take advantage of this opportunity?

In earlier studies, the aforementioned method of applying smooth perturbations was used to change the external force during movement tasks. To apply a similar method during force production tasks in isometric conditions, one has to introduce smooth positional perturbations. This was done using the aforementioned *inverse piano* device (Martin et al., 2011). To remind, in the *inverse piano*, keys press on the fingers and move them up and down with the help of programmable linear motors. This device allows moving the fingers at a controlled speed and over controlled amplitudes while the subject is performing a force production task.

If a person presses on a force sensor, and the sensor suddenly starts to move upward, the force is expected to increase because the agonist muscles (finger flexors) are stretched and muscle force is known to increase with muscle length due to both peripheral tissue properties and spinal reflexes. For small, smooth perturbations, the dependence of force on sensor coordinate is close to linear (Ambike et al., 2016a), which allows computing the intercept and slope of the corresponding regression line. The intercept is analogous to RC for the effector: The effector is expected to move to this coordinate if the external force becomes zero. The slope reflects the apparent stiffness of the effector (k). As such, these two measured parameters are direct reflections of two hypothetical parameters at the control level, the *r*-command and the *c*-command.

Theoretically, for a given force value, there is an infinite number of combinations of {RC; k} pairs that can be used to perform the task of reaching a particular force magnitude, F_{TASK}. A few such combinations are illustrated in Fig. 5.28A. They all satisfy the equation: $F_{TASK} = k \cdot (AC - RC)$, where AC stands for the actual coordinate of the effector, which does not change during isometric force production. Selecting a reference frame, where AC = 0, turns this equation into $F_{TASK} = -k \cdot RC$. In the abundant {RC; k} space, this equation defines a hyperbolic solution space for a given magnitude of F_{TASK} (the UCM).

It is important to emphasize that a subject performing this experiment does not have to use solutions that covary along the UCM. Indeed, the target size for F_{TASK} defines permissible deviations orthogonal to the UCM but not along the UCM. As shown in Fig. 5.29, the same accuracy of performance can be achieved with different patterns of variability of RC and k. For example, only RC can vary while keeping k constant across trials (open circles), or the other way around (filled circles), or even they can covary in the orthogonal to the UCM direction (crosses in Fig. 5.29). This illustration explains that the prediction of {RC; k} variation across trials along the UCM is nontrivial.

To explore this prediction, one has to define RC and k in individual trials, which was done using small smooth positional perturbations of the effectors (individual fingers or the whole hand) using the *inverse piano*. Assuming that the subject in such an experiment does not react to the finger motion, one can measure the force and coordinate changes and use linear regression to identify values of RC (intercept) and k (slope). Repeating this procedure several times for the same background force level produces a family of {RC; k} solutions needed for further analysis. Of course, the assumption of "nonintervention" by the subject is a strong one. It is based on an earlier study that explored human behavior under the instructions "not to intervene"

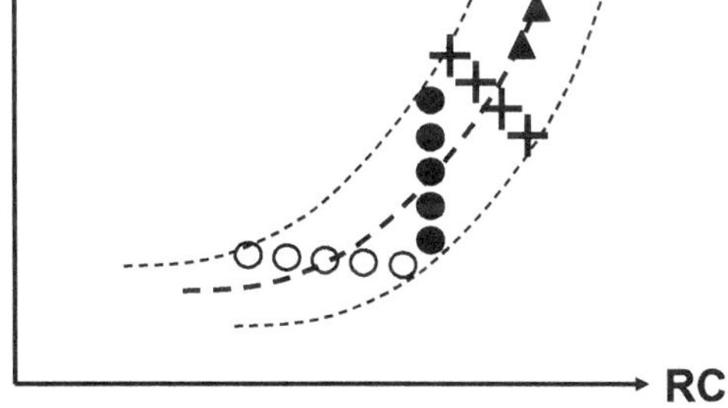

Figure 5.29 The task of constant force production by an effector is associated with a solution space (UCM), a hyperbola in the control space, {RC; k} space. Permissible errors (ΔF) are shown as *thin lines* parallel to the UCM. The same accuracy of performance can be achieved with different patterns of intertrial variability of referent coordinate (RC) and apparent stiffness (k): Only RC can vary while keeping k constant across trials (*open circles*), and k can vary while keeping RC constant across trials (*filled circles*); they can covary in the direction orthogonal to the solution space (UCM) (*crosses*) or along the UCM (*triangles*).

and "correct as quickly as possible" when an external perturbation was applied to an effector (Latash, 1994). The behavior was much more consistent under the former instruction, suggesting that humans are indeed able to ignore small and smooth external perturbations (rather than reacting to them unintentionally).

The nonlinear (hyperbolic) shape of the UCM does not allow applying the analysis of intertrial variance described earlier (Section 4.2). There are two methods that can be used instead. One of them is applying hyperbolic regression to the {RC; k} data pairs collected in different trials. This method is based on assuming a specific covariation between the two control variables.

The other method is somewhat more sophisticated. It was developed in earlier studies of motor tasks when sets of performance variables had to covary across trials to ensure success (Kudo et al., 2000; Müller and Sternad, 2003; Müller and Sternad 2004). For example, throwing a basketball into the basket requires covariation among such elemental mechanical variables as coordinates of release, ball speed at release, and the angle of ball release. Effects of this covariation on accuracy of performance can be estimated by removing the covariation and exploring effects of this manipulation on indices of accuracy (for example, the percentage of successful trials or deviations from the center of the basket). This can be done in a relatively straightforward way: Take values of individual elemental variables from *different trials* (then they would not covary) and compute indices of performance based on rules of mechanics.

For example, three coordinates of the ball at the moment of release can be taken from trials 3, 7, and 11, ball speed from trial 10, and the two-dimensional angle of release from trials 2 and 20. Such surrogate sets of elemental variables can be chosen at random and, as a result, their means and standard deviations (SDs) will be about the same as in the original data set. Performance variables computed for the surrogate sets of elemental variables will be characterized by mean and SD values. If covariation in the original data set helped reach better performance accuracy, SD of the salient performance variable should be larger for the surrogate data set (SD_{SUR}) compared with the actually collected data (SD_{ACT}). For example, one can use the ratio between SD_{SUR} and SD_{ACT} as an index of synergy stabilizing performance.

Recording the {RC; k} pairs over a series of trials with constant force production by a finger or by the four fingers of a hand resulted in data points closely clinging to a parabolic curve corresponding to a constant force magnitude close to F_{TASK} (schematically shown by the filled triangles in Fig. 5.29; actual data are shown with open triangles in Fig. 5.30; Ambike et al., 2016a; Reschechtko and Latash, 2017). The method of computing SD_{SUR} and SD_{ACT} described in the previous paragraph showed SD_{SUR} values at least an order of magnitude higher than SD_{ACT}. These results show that synergies stabilizing salient performance variables can be organized not only at the level of abundant performance variables produced by elements contributing to the task but also at the level of abundant control variables. Given that each action, even with a single effector (e.g., a finger), is controlled with at least two opposing groups of muscles, agonists and antagonists, at the level of control, even a single-effector action is abundant at a control level and can be stabilized by covariation of control variables.

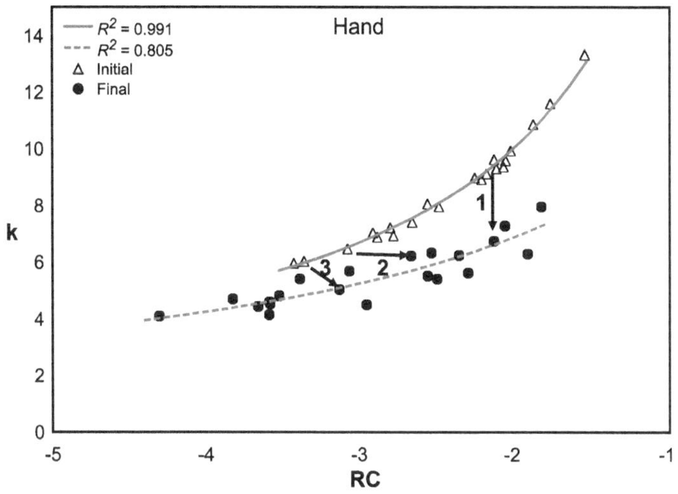

Figure 5.30 Multiple trials at the task of constant force production by the hand are associated with data (*open triangles*) distributed tightly along the solution space (*solid line*, the UCM) in the control space, {RC; k}. After the visual feedback on the hand force was removed, force drifted to lower magnitudes associated with a different UCM (*dashed line*) and the data (*filled circles*) that were not as tightly constrained to the UCM (compare the hyperbolic regression R-values). The three *numbered arrows* represent three scenarios of possible drifts in RC and *k*. Modified by permission from Reschechtko, S., Latash, M.L., 2018. Stability of hand force production: II. Ascending and descending synergies. Journal of Neurophysiology 120 1045–1060. Copyright by American Physiological Society.

5.6.2 Unintentional changes in performance at the referent coordinate level

The method of measuring RC and *k* allows exploring the aforementioned phenomenon of unintentional drifts in performance. Imagine, for example, that a person is producing an accurate force level by pressing with a fingertip on a force sensor with the help of visual feedback on the force level. Then, the feedback is turned off, and the subject is instructed: "continue doing what you have been doing." Under visual feedback, subjects show {RC; *k*} pairs that cling closely to the corresponding UCM, e.g., to one illustrated in Fig. 5.30 with the solid curve. A drift in force to lower magnitudes implies that the UCM on the {RC; *k*} plane drifts downwards as shown with the dashed UCM in Fig. 5.30. Individual data points can drift in different ways showing a drop in RC or in *k* or both (illustrated with arrows in Fig. 5.30).

In several experiments (Ambike et al., 2016b; Reschechtko and Latash, 2017), a small smooth perturbation was applied before the drift, when the subject used visual feedback to produce accurate force, and 10–30 s after turning the visual feedback off when the force drift magnitude was typically large. Repeating this procedure multiple times allows exploring changes in the {RC; *k*} values and in their covariation when the force produced by the effector drops. The experiments have shown that all three scenarios in Fig. 5.30 are possible, although a slow drop in *k* seems to be most consistent. The drift of *k* to lower values was also observed during trials when the subjects maintained constant force with the help of continuous visual feedback. Under this condition, no force drift is observed. The drift in k was coupled to RC adjustments that allowed keeping the force magnitude unchanged. In other words, the cloud of points on the {RC; k} plane slid along the UCM as illustrated in Fig. 5.31 (compare the triangles and circles in this figure). This consistent drop in k suggests a drop in the level of muscle coactivation during long-lasting force production, a phenomenon that will be discussed in more detail in one of the next subsections.

In addition, following the force drift, the individual {RC; k} data points showed larger deviations orthogonal to the UCM reflecting a drop of force stability (illustrated by the two data point distributions with and without visual feedback in Fig. 5.30). The loss in force stability was confirmed also by the more typical analysis within the UCM hypothesis framework in experiments when the subjects performed this task by pressing with the four fingers of a hand and force modes were used as elemental variables (Parsa et al., 2016; Reschechtko and Latash, 2017). Total hand force could be analyzed in those studies as the sum of individual finger forces and also as a consequence of specifying RC and k at the hand level, RC_{HAND} and k_{HAND}. Both analyses suggested strong stabilization of hand force when the subjects had visual feedback on the hand force magnitude, and loss of hand force stability following the hand force drift after the visual feedback had been turned off. Actually, the unintentional force drop might already be viewed as a sign of lost stability, confirmed by the mentioned findings.

5.6.3 Are there descending and ascending synergies?

The concept of {RC; k} pairs at the hand level and at the individual finger level results naturally in a control scheme illustrated in Fig. 5.32. When a person is asked to press

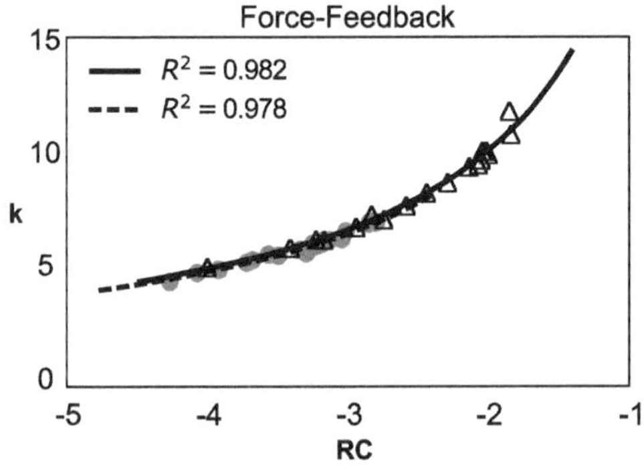

Figure 5.31 This figure is organized similarly to Fig. 5.30. In these trials, however, the subject always had the benefit of visual feedback on total force. As a result, hand force did not drift, but there was a drift of apparent stiffness (k) to lower magnitudes compensated by a drift in RC. Note the consistently high R-values for the hyperbolic regression for both data sets, at the beginning of the trial (*triangles*) and at its end (*circles*).
Modified by permission from Reschechtko, S., Latash, M.L., 2017. Stability of hand force production: I. Hand level control variables and multi-finger synergies. Journal of Neurophysiology 118 3152–3164. Copyright by American Physiological Society.

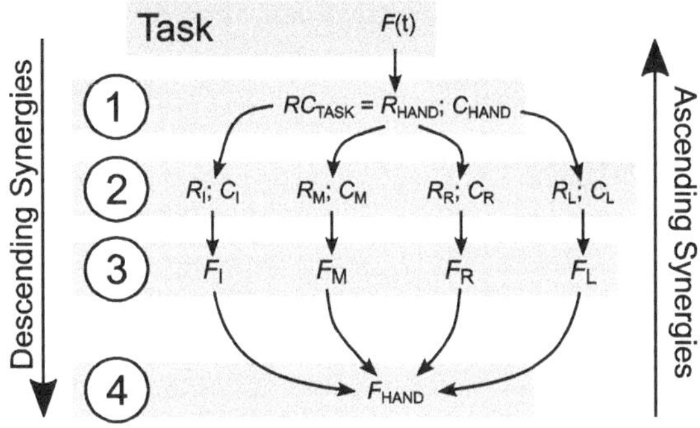

Figure 5.32 A schematic illustration of the notions of descending and ascending synergies within a hierarchical system. At each level, the controller is assumed to vary two basic commands, the *R*-command and the *C*-command (similar to the *r*- and *c*-commands within the equilibrium point hypothesis). *R*-command produces referent coordinate (RC), whereas *C*-command modifies apparent stiffness (k) for the effector.
Reproduced by permission from Reschechtko, S., Latash, M.L., 2018. Stability of hand force production: II. Ascending and descending synergies. Journal of Neurophysiology 120 1045–1060. Copyright by American Physiological Society.

with all four fingers of a hand and produce a certain magnitude of total force (F_{HAND}), we assume that this value is somehow (BMC) represented at the top of the control hierarchy. The desired magnitude of the salient performance variable is used to specify RC values at the hand control level, which can be represented as RC_{HAND} and k_{HAND}. Note that the former is a consequence of the *R*-command (reciprocal command, reflecting unidirectional shifts of RCs for the agonist and antagonist muscle groups), whereas the latter is a consequence of the *C*-command (coactivation command, reflecting counterdirectional shifts of RCs for the agonist and antagonist muscle groups). To involve the individual fingers, the $\{R_{HAND}; C_{HAND}\}$ pair has to be transformed into similar pairs to the index, middle, ring, and little fingers: $\{R_I; C_I\}$, $\{R_M; C_M\}$, $\{R_R; C_R\}$, and $\{R_L; C_L\}$, respectively. The four pairs produce finger forces that, ultimately, sum up to produce actual hand force.

Typical analyses of synergies stabilizing performance described so far have all followed the same direction: from a higher-dimensional space specified at a higher level in Fig. 5.32 to a lower-dimensional space specified at a lower level of that figure. Indeed, F_{HAND} was analyzed as a consequence of specifying a $\{RC_{HAND}; k_{HAND}\}$ pair or as a consequence of generating four individual finger forces; individual finger forces were analyzed as consequences of specifying corresponding pairs $\{RC_i; k_i\}$ to each finger (the subscript i stands for different fingers), and F_{HAND} was also analyzed as a consequence of specifying four $\{RC_i; k_i\}$ pairs to the individual fingers. F_{HAND} was also analyzed as a consequence of specifying commands to individual fingers (finger modes, Danion et al., 2003b) that take into account the nonindependence of finger actions due to multiple factors ranging from peripheral connective tissues and multifinger extrinsic hand muscles to overlapping cortical representations (reviewed in Schieber and Santello, 2004). All these analyses followed the same descending direction in Fig. 5.32 and can be addressed as *descending synergies* (Reschechtko and Latash, 2018).

There is, however, a possibility to perform analysis of synergies in the opposite, ascending, direction and ask the following question: Are {RC_i; k_i} pairs to individual fingers organized to stabilize the {RC_{HAND}; k_{HAND}} pair? Note that the {RC_i; k_i} pairs to individual fingers show much larger intertrial variability orthogonal to the UCM resulting in larger intertrial variability of individual finger forces compared with that of F_{HAND}. This leads, in particular, to multifinger synergies stabilizing F_{HAND} in the finger force space.

Four linear springs connected in parallel form a system with total stiffness equal to the sum of stiffnesses of the individual springs: $k_{HAND} = \sum k_i$, where i = I, M, R, and L. This allows splitting the question about ascending synergies during the hand action into two separate questions. (1) Is there a synergy among k_i stabilizing k_{HAND}? and (2) Is there a synergy among RC_i stabilizing RC_{HAND}? The second question is more complicated to analyze because RCs in a system with parallel fingers do not sum up. So, linear relations between RC_i and RC_{HAND} have to be assumed and discovered for each person individually. This is not a unique situation. For example, in studies of multimuscle synergies stabilizing mechanical variables, links between changes in muscle group activations and in the mechanical variables were discovered using such tool as multiple linear regression (Krishnamoorthy et al., 2003a,b). Of course, using such tools introduces simplifications into the actual system.

Studies of ascending synergies in the four-finger system produced somewhat unexpected results. There were no synergies stabilizing k_{HAND} in the k_i space, but there were synergies stabilizing RC_{HAND} in the RC_i space. Actually, positive intertrial covariation of k_i across fingers tended to destabilize k_{HAND}, likely contributing to the aforementioned drift in k observed frequently during continuous force production. There are two explanations for these results. The first one follows an earlier assumption (Feldman, 2015) that the *R*-command is hierarchically higher than the *C*-command: Changes in the former define desired actions, whereas changes in the latter adjust as required by external conditions and their expected changes. For example, following a fast movement, there is typically an increase in the level of coactivation of the opposing muscles (Gottlieb et al., 1989a). Then, the level of coactivation drops gradually without a change in the final position of the effector suggesting that the *C*-command drops while the *R*-command remains invariant.

Another interpretation is built on the muscle design of the human hand. Individual fingers are moved by multifinger extrinsic muscles (such as FDP, FDS, and extensor digitorum communis). Coactivating these flexor and extensor muscles changes the apparent stiffness coefficients (k_i) of all four fingers in parallel. So, trial-to-trial variation of the *C*-command to one of the fingers is expected to lead to similar variation of this command to the other three fingers of the hand, i.e. to positive covariation of k_i leading to destabilization of their sum (k_{HAND}).

The last part of this section naturally leads to a seemingly naïve question: Why do people (and other animals) coactivate agonist and antagonist muscles? Obviously, if one considers the resultant forces and torques, muscle coactivation is wasteful: It requires more metabolic energy compared with a situation when only agonists are active. We discussed this question briefly in an earlier section, but now we will try to address it in more detail based on the knowledge of {RC; k} synergies stabilizing action.

5.6.4 Why do people coactivate muscles?

As mentioned earlier, all natural movements are produced by an interaction between muscles (muscle groups) with opposing actions commonly addressed as agonists (those that produce forces and torques in the desired direction) and antagonists (those that produce forces and torques in the opposite direction). This classification is not always exact, for example, in cases of biarticular and multiarticular muscles, but acceptable for the purposes of this section. If an animal wished to perform an action by an effector in a certain direction with minimal energy expenditure, the most obvious and frugal strategy would be to activate agonist muscle(s) without involving antagonists. This is not what happens during natural movements: There are frequent episodes of simultaneous activation of muscles with opposing actions addressed as agonist–antagonist coactivation (reviewed in Smith, 1981). The phenomenon of muscle coactivation has been known for over 100 years, starting at least with classical papers by Demeny (1890) and Babinski (1899). Arguably, the first review on this phenomenon was written nearly 100 years ago (Tilney and Pike, 1925).

At the single-joint level, healthy persons show muscle coactivation across various actions ranging from steady-state tasks to quick movement and force production tasks. During fast single-joint actions, both movements and force generation in isometric conditions, muscles commonly show the so-called triphasic activation pattern starting with an increase in the agonist activation at the time of action initiation (Ghez and Gordon, 1987; Gottlieb et al., 1989b; Corcos et al., 1990). During this first phase, the antagonist muscle(s) show an increase in their activation level apparently resisting the action of the agonist(s). The magnitude of this antagonist coactivation increases with parameters of the agonist activation level following changes in such task parameters as action speed and inertial load (Corcos et al., 1989; Gottlieb et al., 1989b). This initial coactivation may serve various purposes, in particular, those related to complexity of muscle action in natural joints and ensuring sufficient joint apparent stiffness (see later and Hasan, 1986). After the end of a quick action, there is usually an elevated level of muscle coactivation, which takes a relatively long time, up to 10 s, to disappear (Gottlieb et al., 1989b; Suzuki and Yamazaki, 2005).

During more natural movements, not limited to single joints, similar coactivation phenomena are observed during fast actions. Analysis of such coactivation patterns may be performed at the level of muscles crossing individual joints or at the level of muscle groups addressed as *muscle synergies* (d'Avella et al., 2003; Ivanenko et al., 2004) or *muscle modes* (Krishnamoorthy et al., 2003a). For example, when a standing person performs a whole-body action, for example, swaying in the anterior–posterior direction or initiating a step, two stable muscle modes with opposing action are commonly seen. One of them unites muscles on the dorsal side of the body, whereas the other mode unites muscles on the ventral side of the body. These modes are typically recruited in a reciprocal fashion: One of them shows an increase in activation level whereas the other is quiescent or shows a drop in its activation level. Alternatively, they can be coactivated with a parallel increase in the activation level of both. Note that this type of coactivation is not joint-specific; it acts at the whole-body level. Low levels of muscle coactivation may be seen during natural standing, and the magnitude of muscle coactivation increases during standing in challenging conditions such as during standing with reduced supporting area or under conditions of low friction between the feet and the supporting surface (e.g., during standing on roller skates) (Berger et al., 1992; Shiratori and Latash, 2000; Krishnamoorthy et al., 2004; Asaka et al., 2008, 2011).

Typically, muscle coactivation is discussed at the level of movement mechanics. Its interpretations focus on effects of coactivation on movement characteristics (e.g., Frysinger et al., 1984; Hirokawa et al., 1991; Tal'nov and Kostiukov, 1991), presumably mediated by changes in the stiffness-like properties of the involved effectors, their *apparent stiffness* (Latash and Zatsiorsky, 1993), defined as the ability to generate resistive forces (moments) per unit of linear (angular) displacement. Activating a muscle leads to a significantly steeper dependence of its force on length or, in other words, to higher apparent stiffness. This effect can be seen in deafferented muscles, i.e., muscles without reflexes. Intact muscles show even higher values of apparent stiffness, and the dependence of muscle force on length characteristic becomes more linear (compare the thick solid and dashed lines in Fig. 5.33A; cf. Nichols and Houk, 1976).

If one considers a joint spanned by several muscles acting in parallel (both agonists and antagonists), in a linear approximation, apparent stiffness of the joint will represent the sum of the apparent stiffness values for the individual muscles. This is illustrated with the thick line in Fig. 5.33B, which is the algebraic sum of the thin individual muscle characteristics. The summation of individual muscle apparent stiffness values is one of the reasons coactivation leads to an increase of the apparent stiffness of joints and multijoint effectors (cf. Nielsen and Kagamihara, 1992). Changes in the apparent stiffness have several mechanical effects including those on movement speed and stability of states and of trajectory.

In particular, speed of movement produced by a muscle group is defined by both changes in neural control variables to those muscles (such as λ for a single muscle, *r*-commend for a joint, or R-command for a multijoint effector, Feldman, 1966, 1980) and by the apparent stiffness of the effector (defined by the *coactivation command* or *c*-command, Feldman, 1980). Indeed, a mass−spring system, in a linear approximation, is characterized by natural frequency: $\omega_0 = \sqrt{k/m}$, where k is stiffness and m is mass (cf. Milner and Cloutier, 1993). Fast movements are only possible in systems with sufficiently high ω_0, which requires high k.

Imagine that you try to move an object attached to one end of a spring by grabbing and moving very fast the other end of that spring. If the spring is very compliant (low k), your action will mainly stretch it, and it will take time for the object to follow the movement of the hand. If the spring is very rigid (high k), the object will follow the hand motion with little delay (which becomes zero if the spring is replaced by a rigid rod). Hence, increasing the apparent stiffness of effectors is a necessary condition for fast movements, which corresponds to many experimental observations of increased coactivation levels for both very fast movements and isometric contractions

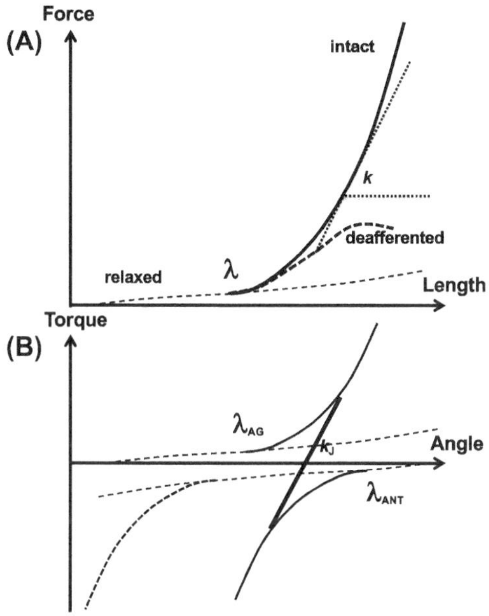

Figure 5.33 A schematic illustration of muscle force–length and torque–angle characteristics. (A): The shallow *dashed line* shows the force–length characteristic of a relaxed muscle. When the muscle is stretched beyond the stretch reflex threshold (λ), its torque–angle characteristic becomes steeper (*thick, solid line*). A muscle without reflexes (deafferented) shows a less linear torque–angle characteristic (*thick dashed curve*). Apparent stiffness (k) is defined as the angle of the tangent of the characteristic with the X-axis. B: If the antagonist muscle shows nonzero coactivation (*solid curve*), the overall joint characteristic (*black, thick line*) becomes steeper as compared with no coactivation (the *dashed antagonist* characteristic). This corresponds to higher apparent stiffness, k_J. Agonist produces positive torque values; antagonist produces negative torque values.
Reproduced by permission from Latash, M.L., 2018. Muscle co-activation: Definitions, mechanisms, and functions. Journal of Neurophysiology 120 88–104. Copyright by American Physiological Society.

(Ghez and Gordon, 1987; Corcos et al., 1989; Bennett et al., 1992). Hasan (1986) explored an optimization approach based on minimizing an "effort" cost and showed that some optimal magnitude of joint apparent stiffness is needed to match experimentally observed trajectories. Studies of comparably fast movements against different inertial loads presented evidence for an increase in coactivation (and apparent stiffness) with the inertial load (Lestienne, 1979) in line with Hasan's conclusions and with the equation for natural frequency presented in the previous paragraph.

Imagine now that you sit in a chair, place a hand to a comfortable location in space, and try to keep it steady. Sometimes, unexpected small forces act on the hand. If the arm muscles are relaxed, these forces will produce relatively large hand deviations. If the muscles show substantial cocontraction, the hand deviations will become smaller. So, muscle cocontraction helps stabilize position of a multijoint kinematic chain.

This is true for effectors, such as joints and multijoint limbs, with a fixed origin. In the previous example, sitting in the chair prevented the trunk from moving and made the body immune to possible small perturbing forces applied to the hand. The situation changes if the whole body is free to move.

For example, imagine that you perform the same experiment (holding a hand position in space) while standing and unexpected force perturbations are applied to the hand. When you stand on the floor, the feet are not glued to it and, as a result, effects of muscle coactivation in the arm and along the vertical body axis on responses to external force perturbations become complex and potentially detrimental for stability of the whole-body posture. If all the joints are very rigid (high agonist–antagonist muscle cocoactivation), forces applied to the hand are readily transmitted to other body segments and act to destabilize the body. Here, we define postural stability as an ability to stand, without falling or making a step, under brief external force perturbations. So, coactivation does not look like a good strategy to ensure stability of vertical posture or of any multijoint effector with the origin that is free to move. This general conclusion has been recently supported in a study showing worse indices of postural stability in young healthy persons during standing with purposefully increased cocontraction of muscle groups in the lower extremities (Yamagata et al., 2019a).

Indeed, consider a long log placed on one of its flat ends. This object is very rigid and very unstable: A brief force pulse applied to its upper end can easily tip it over. Note that if the other end of the log were glued to the floor, it would show very high resistance to external perturbations (high stability). Patients with Parkinson's disease are known for their problems with postural stability. They stand in a rigid way, particularly at advanced stages of the disease, and show very small spontaneous deviations of the body (small postural sway, Horak et al., 1992). However, even a small external force pulse can make them fall over or make a sequence of small steps that frequently fail to prevent a fall. In contrast, persons who practice T'ai Chi typically stand with their muscles relaxed and show better postural stability compared with the general population (e.g., Gatts, 2008).

As the previous paragraphs suggests, muscle coactivation may not be such a good idea from the point of view of postural stability in the field of gravity. Nevertheless, humans (and other animals) do coactivate muscle while standing and show elevated levels of agonist–antagonist coactivation when standing in challenging conditions. What could be the reason for this, apparently suboptimal, strategy?

An answer has been suggested recently related to the idea of action stability organized at the level of control variables to the involved muscle groups (Latash, 2018b). Consider the control of a single effector. As mentioned earlier (see Section 5.2), if one of the muscles within an agonist–antagonist pair is quiescent, changing λ to that muscle within the subthreshold range formally is associated with coadjusted changes in both *r*- and *c*-commands, but it has no effect on behavior. As illustrated in an earlier figure (Fig. 5.8), as long as the control signal to the antagonist muscle, λ_{ANT}, stays at values subthreshold for muscle activation, any changes in λ_{ANT} would lead to no behavioral effects. As a result, the only method to control behavior is to change the control variable to the agonist muscle, λ_{AG}. In other words, the two-dimensional control space is reduced to a single-dimensional space. This makes the control degenerate, the number of variables manipulated at the control level becomes nonabundant (one), and this does not allow using control-level synergies stabilizing behavior. Keeping both λs within the activation range makes it possible to vary two variables at the control level (λ_{AG} and λ_{ANT}, or *r*- and *c*-commands), independently of each other, and thus organize stability of salient task-specific variables. Such {RC; *k*} synergies depend crucially on the possibility to modify both R-command and C-command to modify actions (Ambike et al., 2016a, 2018; Reschechtko and Latash, 2017, 2018; see Section 5.6.1). This becomes impossible if the C-command is zero or negative.

Based on the idea of {RC; *k*} synergies stabilizing salient performance variables, we can assume that one of the functions of muscle coactivation, and maybe its main function, is to ensure abundance of variables at the control level that allows stabilizing performance. Why do healthy persons coactivate more in challenging conditions? The answer is straightforward. By definition, a challenging motor task is a task associated with poorly predictable changes in external conditions and/or intrinsic body states. This means that such tasks are more likely to face unexpected perturbations and require corrective actions; in addition, such corrective actions are likely to deal with stronger destabilizing effects. The destabilizing factors, such as external force perturbations, increase chances of large, unpredictable changes in position of the effectors that can move joints to zones characterized by zero coactivation, thus making control at the {*r*; *c*} level impossible and forcing the degenerate control with only one control variable able to modify the ongoing action. Increasing agonist–antagonist coactivation effectively increases the size of the spatial range where both opposing muscle groups are within the activation zone. Therefore, unpredictable changes in muscle length are less likely to drive either of them outside this zone preserving the abundant control and contributing to action stability. Of course, direct effects of coactivation mediated by changes in apparent stiffness of the effectors are also likely to be important: As mentioned earlier, in cases of kinematic chains with fixed origins, higher coactivation improves stability due to direct mechanical effects of increasing apparent stiffness of the joints.

There is one more potential beneficial effect of muscle coactivation when the task is performed in conditions of possible strong external force perturbations. Such perturbations require quick corrective actions mediated by large-magnitude shifts of control variables (such as λ for single muscles). Imagine that an external perturbation requires a quick action produced by the antagonist muscle(s), i.e., it requires moving λ for the antagonist muscle(s) deep into the activation zone. We should also consider the fact that signals within the central nervous system cannot change instantaneously. Estimates of the highest rate of λ change in single-joint movements performed under the instruction to move "as fast as possible" produced values on the order of 800°/s. Although this is a very high speed, it can contribute substantially to reaction time if λ for the antagonist muscle is outside its activation range (as in the previous example illustrated in Fig. 5.8). For example, if λ_{ANT} is 40° away from the current muscle length, it would take the central nervous system an extra 50 ms to initiate muscle activation as compared with a situation with λ_{ANT} equal to the current muscle length. This is potentially a substantial addition to the total reaction time, which can make a difference between success and failure in dealing with perturbations.

As a transition to the next section, let me also mention the phenomenon of increased agonist–antagonist coactivation in persons with various movement impairments, from healthy older persons to persons with atypical development and to patients with neurological disorders (Aruin et al., 1996; Lee and Ashton-Miller, 2011; Arias et al., 2012; Richards and Malouin, 2013; Hirai et al., 2015; Kitatani et al., 2016). Increased coactivation seems to be a universal phenomenon unrelated to specific problems leading to movement discoordination and other disorders. Within the suggested scheme, one reason for this phenomenon may be that persons with motor impairments perceive everyday tasks as challenging, leading to patterns of muscle activation similar to those observed in healthy persons in challenging situations. As a result, they facilitate patterns of muscle activation that ensure abundance of control variables and allow using control-level synergies stabilizing salient performance variables. It is also possible, of course, that coactivation patterns are produced by mechanisms only indirectly related to the production of voluntary movements, e.g., those reflecting reorganization of spinal mechanisms typical of spasticity (see the next section).

5.7 Disordered control with referent coordinates

Before discussing what can go wrong with the introduced scheme of control with RCs, let me reiterate that changing RCs should be viewed as a set of tools used by the brain to achieve motor goals. The brain does not compute requisite RC(t) time profiles (or anything else!) based on the desired forces and trajectories but uses memory to generate RC(t) functions for specific tasks. To use a metaphor, when a person drives a car, he or she presses on the gas and brake pedals and turns the steering wheel to produce a desired car trajectory. This statement does not mean that the driver's brain computes how specific pressure on the gas pedal would lead to a change in fuel injection into the cylinders, motion in the engine, transmission of this motion to the wheels, and ultimately interaction between the wheels and the pavement leading to car motion. There are also complex transformations between pressure on the brake pedal and friction between the wheel surface and brake pads and between effort applied to the steering wheel and turning of the car wheels. When we learn how to drive a car, we develop and memorize mappings between our actions on the pedals and steering wheel and their effects on car motion. These memories allow generalization to different driving situations and different cars, although, as every driver knows, it takes some time to recalibrate one's actions and adjust to a new car.

Along similar lines, during childhood, we develop mappings between our sets of tools, RC(t) functions, and motor performance in typical situations. The RC(t) functions can be used in a feed-forward manner if a quick action is needed and can be adjusted based on sensory signals if accuracy is important or in cases of unexpected changes in salient performance variables (such as trajectories of effectors) that require corrections.

With this in mind, consider what kind of problems can emerge in the neural control of movements with pathological changes in different components involved in the control with RC(t) time functions. Tools can be used wrongly, for example, if the actor does not have a stable, veridical representation of the external space (e.g., when wearing prismatic glasses that distort visual signals) or when external forces and their changes cannot be predicted (e.g., when trying to make fast movements under water for the first time). Tools can show drifts unrelated to one's intention, thus leading to inaccurate movements (examples of such behaviors were discussed in Section 5.4). Tools may be limited in the range of their use (we will discuss the example of spasticity later in this section). Finally, tools can be broken altogether (e.g., in cases of a loss in the natural feedback from proprioceptors addressed as deafferentation). We are not going to consider major peripheral changes in the motor apparatus such as those in cases of amputation or myopathies leading to a decreased ability, or even inability, of muscles to produce force, when aspects of the motor function have to be replaced with external means such as a wheelchair, an exoskeleton, or another person.

5.7.1 Movements after loss of sensory feedback

Loss of proprioceptive feedback is a major disorder, which does not allow the central nervous system to use the method of specifying thresholds of the stretch reflex to the involved muscles because the reflex is absent. An alternative is to specify the output of the alpha-motoneuronal pools directly. This effectively switches the controller to another mode, which cannot take advantage of the adaptive behavior afforded by the stretch reflex loops. This mode of control, however, is compatible with the idea that the central controller specifies parameters of the muscles that define interactions between the effectors and external forces. Indeed, a level of activation of alpha-motoneurons defines a level of muscle activation, and for a given level of activation, muscle force is a function of muscle length. Changing muscle activation leads to changes in the force–length dependence, which effectively translates into a change in the slope of this dependence (see Fig. 5.34) in contrast to the control with RC (λ), which shifts the force–length dependence parallel to the length axis.

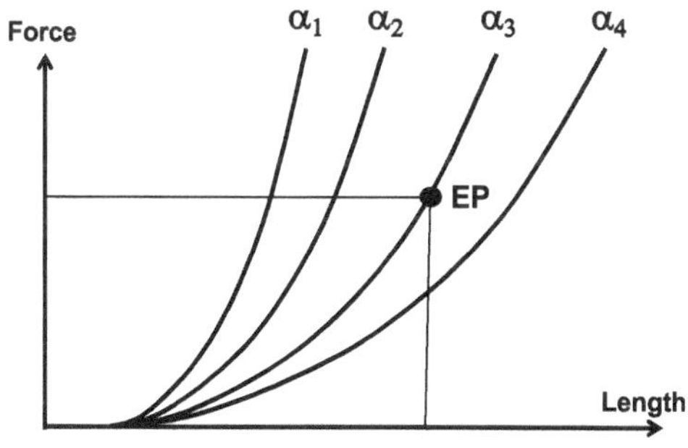

Figure 5.34 An illustration of a set of muscle force–length characteristics for a deafferented muscle. The control variable is the level of activation of α-motoneurons resulting in a certain level of muscle activation. Changing this variable effectively changes the slope of the force–length dependence (cf. $\alpha_1, \ldots, \alpha_4$). For a given α (e.g., α_3) and a given external load, an equilibrium point (EP) is achieved.

Some of the features of control in the intact body are preserved in animals without proprioceptive feedback as demonstrated in experiments on deafferented monkeys performed by the group of Emilio Bizzi (Polit and Bizzi, 1978, 1979; Bizzi et al., 1982). In those experiments, the animals learned fast arm movements into a target under visual control and then performed them in conditions when they could not see the moving arm anymore. After deafferentation, monkeys showed the phenomenon of *equifinality*: When the learned arm movement was suddenly stopped by external force and then released, the arm landed in the target, although the monkeys could not receive information from the perturbed limb and, therefore, were not expected to correct the ongoing neural control process. Note that this result is naturally expected if the controller specifies parameters that define equilibrium states of the effector–environment system, and it is not expected if the controller specifies time profiles of forces generated by the muscles. In another experiment, external force was used to bring the arm into the target very quickly and then to release it. After the release, the arm moved back, *away from the target*, and then reversed its movement and landed in the target again. This result shows that the final equilibrium state of the arm in the original, undisturbed force field was not specified by the controller instantaneously but achieved as a result of a continuous process specifying a sequence of equilibrium states in time addressed as *equilibrium trajectory*.

The mentioned experiments on deafferented monkeys were summarized into a scheme, which is now addressed as the *alpha-model* within the EP hypothesis (in contrast to the original lambda-model introduced by Feldman in 1966). Within the alpha-model, the central nervous system specifies a magnitude of the output of an alpha-motoneuronal pool. This signal defines a force–length dependence for the muscle innervated by that pool (see Fig. 5.35). The muscle comes to an equilibrium state (an EP) when its force equals the external load force. To make a movement from an initial equilibrium state (EP_0) to a desired target, the controller changes the α signal, which can lead to different shifts of the EP depending on the external load characteristic (three such characteristics are illustrated in Fig. 5.35 corresponding to isotonic, isometric, and elastic loads).

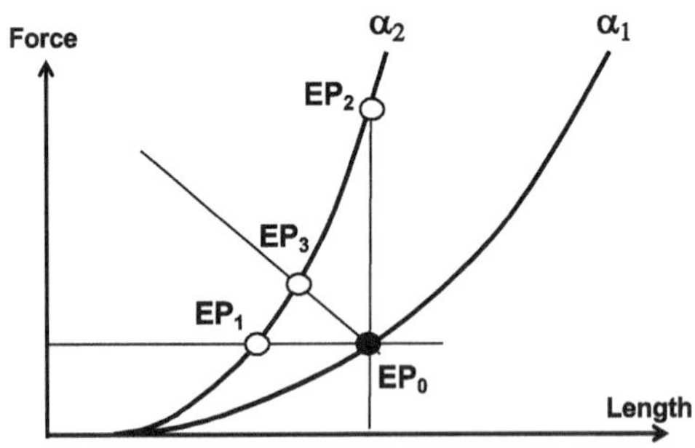

Figure 5.35 Within the alpha-model, voluntary movements are produced by a change in the level of activation of α-motoneurons, for example, from α_1 to α_2. Depending on the external load characteristic (*thin lines*), this can lead to a movement (from EP_0 and EP_1), to force production (from EP_0 and EP_2), or to a combination of the two (from EP_0 and EP_3).

Although the alpha-model allows certain features of voluntary movements to be preserved in deafferented animals, it has to act in the absence of a very important contributor to the force–length characteristic of intact muscles, the stretch reflex (see Nichols and Houk, 1976). In addition, the lack of proprioception disrupts the ability to correct actions if they are perturbed unexpectedly and to coordinate actions across joints of multijoint effectors (cf. Sainburg et al., 1993, 1995).

Problems with motor coordination following partial or complete loss of transmission of signals from proprioceptors have been known for a long time, in particular from clinical observations of patients with advanced stages of syphilis leading to the condition known as *tabes dorsalis*. Such patients had been relatively common in the 19th and early 20th century before effective treatment of syphilis became available. In patients with *tabes dorsalis*, transmission of action potentials along the posterior columns of the spinal cord was interrupted; these columns contain ascending pathways carrying proprioceptive information into supraspinal structures. As a result, patients with *tabes dorsalis* were severely impaired in multijoint actions such as standing upright. In particular, they were unable to stand with eyes closed.

There is a rare condition associated with loss of conduction along peripheral sensory fibers, particularly the large-diameter ones. Patients with this condition (large-fiber peripheral neuropathy) are sometimes imprecisely referred to as *deafferented persons*. Although there is no visible drop in muscle strength in these persons, they show profound problems with motor coordination, in particular, when no help from the visual system is available. A series of studies have shown that persons with this condition have severe problems with interjoint coordination during arm movements such as those imitating cutting a loaf of bread (Sainburg et al., 1993, 1995). The authors interpreted these results as pointing at the importance of proprioceptive information for predicting and accounting for joint interaction torques, i.e., torques in joints produced by motion in other joints of the limb (see Zatsiorsky, 2002). Although this interpretation does not go beyond the level of movement mechanics, it does point at a potentially important role of proprioceptive information in transformations involving task-level (arm-level) neural control variables, such as the *R*- and *C*-commands to joint-level control variables, such as the *r*- and *c*-commands (see Section 5.2).

Note that, despite the lack of stretch reflex in deafferented animals, joint-level control can be based on a pair of variables similar to the *r*- and *c*-commands introduced for the control of joints in healthy animals. Indeed, within the alpha-model, changes in the control signal to two muscles crossing a joint can lead to changes in equilibrium joint angle and joint apparent stiffness. Effects of a change in the *c*-command are illustrated in Fig. 5.36. This figure is organized similarly to Fig. 5.7 with one of the muscles (agonist) producing positive moment of force (or joint torque) and the other muscle (antagonist) producing negative moment of force. Fig. 5.36, for simplicity, assumed symmetrical muscle properties, which is, of course, a major deviation from reality.

Note that changes in the commands to the muscles lead to changes in the profile of their T(α) dependence, where T stands for torque and α for joint angle. For any given joint angle, an increase in the level of muscle activation ($\alpha_3 > \alpha_2 > \alpha_1$ in Fig. 5.36) leads to larger force level (and correspondingly larger torque) and larger apparent stiffness of the muscle. If commands to the two muscles change symmetrically, e.g., from

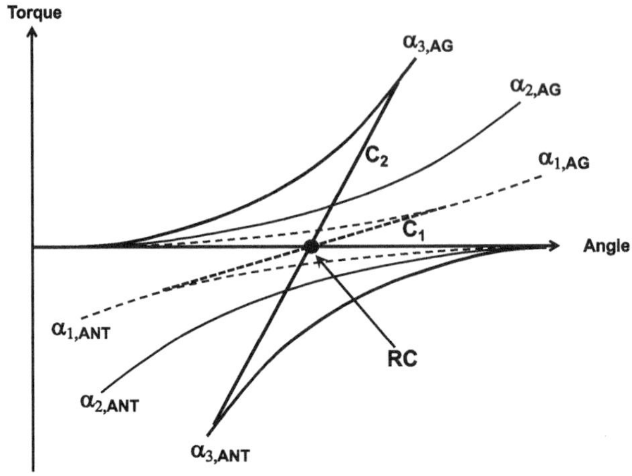

Figure 5.36 Torque–angle characteristics for two muscles crossing a joint, agonist (AG, produces positive torque values) and antagonist (ANT, produces negative torque values), within the alpha-model. Parallel changes in commands to the two muscles, from $\alpha_{1,AG}$ to $\alpha_{3,AG}$ and from $\alpha_{1,ANT}$ to $\alpha_{3,ANT}$, lead to a change in the slope of the joint torque–angle characteristic (cf. the *dashed and solid straight lines*). This control may be described as a change in the coactivation command, *C*-command.

α_1 to α_3, the overall joint characteristic (shown with straight lines) rotates without a major change in its intercept with the *X*-axis. The intercept is marked as RC because it corresponds to the joint equilibrium state in the absence of external torque. So, mechanical effects of a parallel change in the alpha-commands to the two opposing muscles are equivalent to those of the *c*-command within the lambda-model. Within the alpha-model, the *c*-command may be associated with $\min(\alpha_{AG}, \alpha_{ANT})$, which is commonly equal to α_{ANT}.

The situation becomes more complex, if commands to the two muscles change in a reciprocal way: One of them leads to an increase in the agonist muscle activation level, whereas the other command reduces the level of activation of the antagonist. This situation is illustrated in Fig. 5.37. The initial commands to the two muscles are assumed to be $\alpha_{0,AG}$ and $\alpha_{0,ANT}$. Then the command to the agonist increases to $\alpha_{1,AG}$, whereas the command to the antagonist decreases to $\alpha_{1,ANT}$. Note that the joint characteristic in this case shows both translation along the *X*-axis and rotation. So, within the alpha-model, an equivalent of the *r*-command does not only lead to a change in the joint RC (RC_0 to RC_1 in Fig. 5.37) but also to a change in its apparent stiffness (k_0 to k_1). In addition, the joint characteristic becomes nonlinear, even under the assumption of muscles with symmetrical mechanical properties.

The interaction between the control of joint equilibrium position and its apparent stiffness may be one of the factors complicating the control of movements by deafferented animals (including persons). Note that healthy persons can naturally "stiffen" a joint or an effector (e.g., an arm) without moving it, even without looking at that effector. This ability is likely to be compromised in cases of loss of somatosensation from the effector.

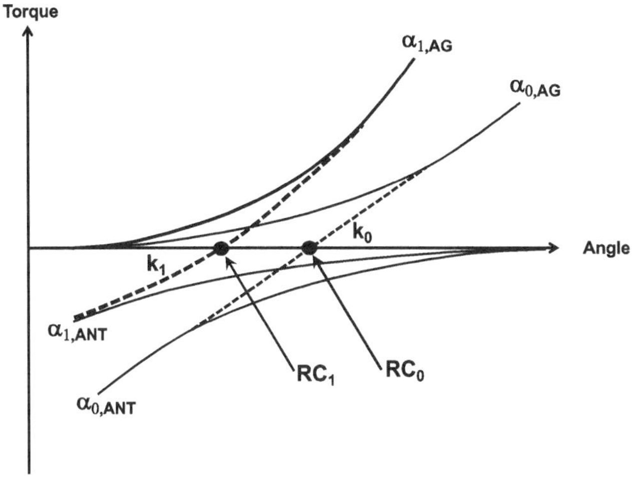

Figure 5.37 Torque—angle characteristics for two muscles crossing a joint, agonist (AG, produces positive torque values) and antagonist (ANT, produces negative torque values), within the alpha-model. An increase in one of the commands, from $\alpha_{0,AG}$ to $\alpha_{1,AG}$, and a decrease in the other command, from $\alpha_{0,ANT}$ to $\alpha_{1,ANT}$, lead to a change in the intercept of the joint torque—angle characteristic (from RC_0 to RC_1) coupled to a change in the slope of the characteristic (from k_0 to k_1). This control may be described as a change in the reciprocal command, *R*-command. Note a change in the shape of the torque—angle characteristic, which becomes less linear.

Action stability can also suffer in conditions with loss of conduction along sensory (afferent) fibers. In particular, stretch reflex has been discussed as an important mechanism stabilizing equilibrium states of effectors (Latash, 2010). The lack of this mechanism may be expected to lead to impaired stability of actions. In addition, sensory feedback loops originating from proprioceptors may be important for stabilization of action as suggested by Martin et al. (2009).

A recent study of a person with large-fiber peripheral neuropathy has shown both differences from and similarities to typical behaviors observed in healthy persons (Cuadra et al., 2019). In that study, the participant was asked to perform a set of single-finger and multifinger tasks. When the participant tried to produce force with one finger at a time (helped by visual feedback on that finger's force), other fingers of the hand produced force unintentionally. This is a well-known phenomenon, addressed as lack of finger individuation or finger enslaving (Kilbreath and Gandevia, 1994; Zatsiorsky et al., 2002; Lang and Schieber, 2003). The person without afferent feedback from the hand showed very large indices of enslaving when she tried to press with one of the lateral (index or little) fingers. This is highly unusual because the index finger typically is the best controlled finger and shows very low enslaving. This observation suggests that the patterns of finger individuation are likely to reflect one's experience with using the fingers in everyday tasks (cf. Slobounov et al., 2002; Schieber and Santello, 2004).

Indeed, loss of somatosensory information makes prehensile tasks, such as taking a sip from a glass without spilling its contents, very challenging. Note that the central fingers, the middle and ring fingers, are responsible primarily for resultant force production, whereas the lateral fingers, the index and little fingers, are specialized for rotational action (Zatsiorsky et al., 2003; Zatsiorsky and Latash, 2008). This suggests a strategy of overgripping the handheld object when it has to be turned, thus increasing the rotational apparent stiffness of the hand with the object and decreasing kinematic consequences of possible errors in the applied moment of force. This can be achieved by coupling commands to lateral fingers (to turn the object) to commands to central fingers (to grip the object), resulting in large enslaving indices generated by lateral finger action in the middle finger forces. No symmetrical effects from central to lateral fingers are expected, and in fact, pressing with the middle and ring finger in that patient led to minor effects on the index and little finger forces.

The patient was able to show basically unchanged indices of multifinger synergies during steady-state force production by the hand and unchanged attenuation of those synergies in preparation to quick force production (anticipatory synergy adjustments, ASAs; see Section 4.3.2). Of course, all these tasks were performed under continuous visual feedback on the total hand force corroborating the earlier conclusions on the importance of visual feedback, not somatosensory feedback, for multifinger synergies (see Sections 5.4.1 and 5.6.2). When the visual feedback was removed, the patient showed a drift of the force to lower magnitudes, on the order of 20% over 15 s, which is comparable with observations in healthy persons (Vaillancourt and Russell, 2002; Ambike et al., 2015a). The force drift suggests that the alpha-command to the agonist muscles drifted unintentionally to lower magnitudes. This conclusion fits the general idea that any natural system tends to drift toward states with lower potential energy independently of the mechanism of the neural control of muscles.

5.7.2 Spasticity

The term *spasticity* refers to a group of pathological conditions characterized by common features such as impaired voluntary control of the affected muscles combined with unintentional muscle contractions, which can be steady-state, jerky, or rhythmical. Such combinations of clinical features can be seen in persons with various neurological disorders including spinal cord injury, multiple sclerosis, cerebral palsy, and stroke. Arguably, the first definition of spasticity was suggested by a great British neurologist, Hughlings Jackson (1889). He defined it as a combination of positive and negative signs. The positive signs ("positive" in this context does not mean "good"!) included excessive muscle activation, as compared with healthy persons, resulting in increased steady contractions, spasms, and involuntary rhythmical activations. The negative signs involved loss of voluntary muscle activation, poor coordination, excessive fatigue, etc. Hughlings Jackson viewed positive and negative signs of spasticity as independent and predicted that alleviating one group of signs was not expected to help the other group, an opinion dominating the field for a long time (e.g., Landau, 1974).

An attempt to link spasticity to neurophysiological mechanisms was made by an Australian neurologist, Lance (1980), who defined spasticity as a state with hyperexcitable stretch reflexes, including tendon jerks. Although this definition was appealing, a few studies quantified the gain of the stretch reflex in spastic animals and reported no changes as compared with healthy animals (Dietz et al., 1981; Dietz, 2000). Besides, this definition addressed only excessive involuntary muscle activation typical of spasticity, not problems with voluntary movements.

Recently, spasticity has been considered within the idea of movement control with changes in RCs for the involved muscles (Levin and Feldman, 1994; Jobin and Levin, 2000; Mullick et al., 2013). The gist of the idea is illustrated in Fig. 5.38. Consider a healthy muscle controlled by changes in λ—threshold of the stretch reflex (Fig. 5.38A). On the one hand, to relax a muscle when its length is close to the upper limit allowed by biomechanical factors (the biomechanical range in Fig. 5.38A),

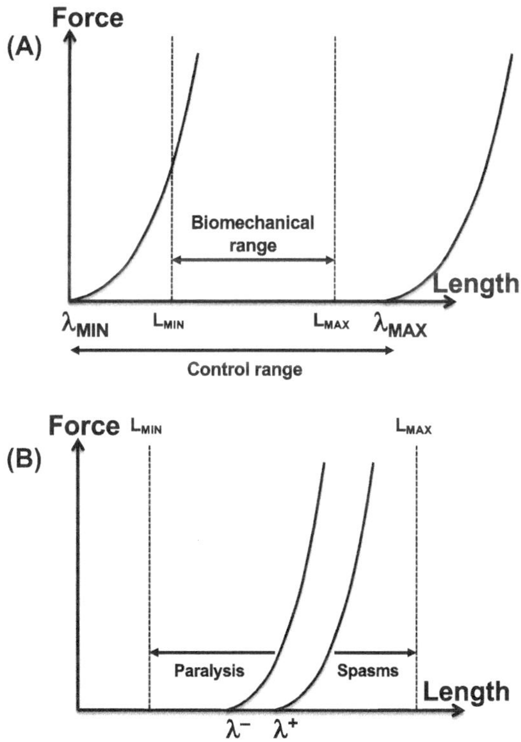

Figure 5.38 (A): In a healthy muscle, the control variable (λ) has to be shifted beyond the biomechanically possible changes in muscle length, between λ_{MIN} and λ_{MAX}, to allow relaxing the muscle and producing large muscle forces within the whole biomechanical range. (B): In a spastic muscle, the range of voluntary shifts of λ is limited, from λ^- to λ^+. This leads to paralysis within a range of muscle length values and uncontrolled contractions within another range (shown with *arrows*).

Modified by permission from Latash, M.L., Zatsiorsky, V.M., 2016. Biomechanics and Motor Control: Defining Central Concepts. Academic Press: New York, NY. Copyright by Elsevier, Inc.

the value of λ has to be moved to the right beyond that limit (λ_{MAX}). On the other hand, to produce large muscle force when it is very short—close to the left limit of the biomechanical range—λ has to be moved to the left beyond that limit (λ_{MIN}). This scheme shows that, for healthy muscles, the neural controller has to be able to shift λ within a wide range, much wider than the biomechanical range of muscle length.

Imagine now what could happen if the ability to shift λ were limited because of an injury to the central nervous system (Fig. 5.38B). If the range of voluntary λ shifts is limited by λ^+ and λ^-, the person would be unable to produce voluntary muscle activations in an area that requires moving λ to the left of λ^- (marked as *paralysis* in Fig. 5.38B), while the muscle would be always active in an area to the right of λ^+ (marked as *spasms* in Fig. 5.38B). Restricted ranges of voluntary shifts of λ in spastic muscles have been documented in studies of stroke survivors with spasticity (Mullick et al., 2013) in support of the scheme illustrated in Fig. 5.38B. Further development of this scheme linked it to motor learning and recovery of motor function in patients after stroke (Turpin et al., 2017; Subramanian et al., 2018).

The idea illustrated in Fig. 5.38B links the *positive* and *negative* signs of spasticity into a single scheme in contrast to the original thinking of Hughlings Jackson. Note that this idea had been challenged earlier in studies of spasticity in patients with spinal cord injury and multiple sclerosis treated with intrathecal delivery of baclofen (a precursor of a presynaptic inhibitor, GABA) (Corcos et al., 1986; Penn et al., 1989; Latash et al., 1990; Latash and Penn, 1996). Those studies showed that alleviating involuntary contractions could unmask more normal-looking movements by the affected extremities.

The idea that spasticity represents a deficit in voluntary control of λ for the involved muscles represents one of the very few examples of using a theory within the field of motor control to address a major clinical problem. The scheme in Fig. 5.38B is clearly incomplete; it addresses only some of the typical problems associated with spasticity. For example, this scheme does not address involuntary spasms triggered by sensory stimuli (for example, a touch to a body area) that can be seen in some patients, rhythmical involuntary actions such as clonus (a cyclical joint movement at 6–7 Hz), and some other components of the typical clinical picture of spasticity. Involuntary spasms can probably be interpreted within the same scheme assuming that some sensory stimuli can move λ into the muscle activation zone by themselves, without subject's intention. Clonus is likely a reflection of autooscillations in the monosynaptic stretch reflex loop (Hömberg et al., 1987; Latash et al., 1989), and therefore, its mechanism may be beyond the scheme considered in Fig. 5.36.

5.7.3 Impaired control of stability (ICS syndrome)

The idea of control with RCs coupled with the apparently redundant design of human effectors leads naturally to the concept of controlled stability of salient performance variables produces by multielement systems (see Section 4.2). The word "*controlled*" in the previous phrase is important; it implies that the central nervous system can stabilize or destabilize a performance variable, weaker or stronger, depending on the planned action. This ability is reflected, in particular, in the phenomenon of ASAs

(Section 4.3.2), a controlled drop in the index of stability of a performance variable typically seen 300 ± 100 ms before the initiation of an action that requires a quick change in that variable.

Although neurophysiological mechanisms of stability are mostly unknown, they are likely to be distributed among different structures within the central nervous system, from the system of Renshaw cells and spinal reflexes, which typically mediate negative feedback effects, to loops involving subcortical structures such as the cerebellum and the basal ganglia (Latash and Huang, 2015; see Section 4.3.3). Injuries to any of those structures can potentially lead to problems with controlled stability of performance, which may include two components. First, stability during steady-state actions may be impaired. Second, ability to generate ASAs before planned quick actions may be impaired, referred to as *impaired agility*. Taken together, the two impairments have been referred to as an impaired control of stability (ICS) syndrome (Latash and Huang, 2015). As described later in this section, many conditions characterized by motor impairments display both impaired stability and impaired agility. There are, however, observations of impaired agility with no change in stability and of differential effects of treatment on these two components of ICS, which justify the term *ICS syndrome*.

Problems in both stability and agility can be seen in populations with very mild changes in motor coordination, such as healthy older persons. Aging is associated with death of neurons at different levels of the central nervous system, from spinal alpha-motoneurons to cortical neurons (Booth et al., 1994; Brooks and Faulkner, 1994; Eisen et al., 1996; Erim et al., 1999; Dinse, 2006). This is accompanied by plastic changes within the central nervous system, changes in properties of muscle and peripheral sensory receptors, and changes in aspects of motor behavior (reviewed in Welford, 1984; Grabiner and Enoka, 1995).

There are reports on impaired multidigit synergies stabilizing salient performance variables, reflected in lower stability indices, during a variety of actions by older persons (Shim et al., 2004; Shinohara et al., 2004; Olafsdottir et al., 2007a; Kapur et al., 2010; Gorniak et al., 2011). Older persons are particularly impaired in their ability to stabilize the rotational actions of the hand and the direction of the resultant force vector produced by the hand (Olafsdottir et al., 2007a; Kapur et al., 2010). At the same time, older persons show significantly shorter and smaller ASAs in preparation to a quick action, compared with younger persons, i.e., impaired agility (Olafsdottir et al., 2007b, 2008).

Although most age-related changes are detrimental to everyday motor behavior, there is one notable exception: Indices of finger enslaving are lower (finger individuation is better!) in older adults compared with healthy young persons (Shinohara et al., 2003a,b). This unexpected observation led to the "*back-to-elements*" hypothesis, which suggests that the death of neurons in pre-M1 brain areas disrupts synergic control and forces the brain to switch to the control of actions at the level of elements—similar to the control of movements in babies who have not yet developed typical synergies for everyday actions (Kapur et al., 2010; Gorniak et al., 2011). This scheme is able to account for the combination of impaired synergic control and better, more individuated, control of individual finger actions in older persons.

Studies in persons with subcortical disorders including Parkinson's disease, olivopontocerebellar atrophy (OPCA, also addressed as multisystem atrophy), and multiple sclerosis have documented significant impairments in both stability and agility across a range of tasks (Park et al., 2012; 2013; Jo et al., 2015, 2017; Falaki et al., 2016). These observations suggest that hypothetical transformations involved in the generation of RCs at lower hierarchical level (for the involved elements) based on the RCs at the task level are likely to depend crucially on the functioning of loops via the basal ganglia (affected in Parkinson's disease) and the cerebellum (affected in OPCA).

Changes in the indices of stability and agility were observed in patients with Parkinson's disease even in the absence of obvious clinical impairments. For example, stage-I of this disorder (after Hoehn and Yahr, 1967) is defined as a stage when clinical signs of the disease are seen in only one half of the body. In such patients, however, similar significant changes in the indices of stability and agility were documented during tasks performed by the hand at the apparently affected side and during similar tasks performed by the hand at the apparently unaffected side of the body (Fig. 5.39; Park et al., 2014). Note the significantly lower baseline indices of multifinger synergies stabilizing total force in the group of patients with Parkinson's disease compared with age-matched healthy control subjects. When the subjects were asked to produce a quick force pulse, the control group showed a consistent drop in the synergy index starting about 250–300 ms before the moment of force change (ASA). In contrast, the patients showed much less consistent ASAs, smaller in both duration and magnitude. This was true for both dominant and nondominant hands. Interestingly, the left (nondominant) hand had a significant advantage reflected in the higher magnitude of the synergy index in both groups. This advantage demonstrated by the nondominant hand could be expected from the dynamic dominance hypothesis (Sainburg, 2005), which postulates specialization of this hand and the corresponding hemisphere for steady-state tasks.

Stage-II of Parkinson's disease is defined as a stage when clinical signs are seen in both sides of the body, but there are no clinical signs of postural instability. Studies of such patients, however, have documented impaired multimuscle synergies stabilizing coordinate of the COP, an important variable for the control of vertical posture (see Winter et al., 1996; Latash and Zatsiorsky, 2016), as well as reduced indices of ASAs, a reflection of impaired agility, in preparation to whole-body actions (Falaki et al., 2016).

Arguably the most convincing example of high sensitivity of the indices of stability and agility to subcortical disorders has been reported in a study of healthy welders (Lewis et al., 2016). Welding is associated with breathing metal vapors, including manganese, which can accumulate in different brain structures, including the basal ganglia, and, after many years of exposure, lead to clinical signs somewhat similar to those in Parkinson's disease (Bowler et al., 2007). The mentioned study of healthy professional welders has shown significantly reduced indices of stability during multi-digit tasks, and these indices correlated with MRI indices in the globus pallidus, a structure within the basal ganglia.

Indices of stability and agility are sensitive to drug treatment: both improve in patients with Parkinson's disease after taking a dopamine-replacement medication (Park et al., 2014; Falaki et al., 2017). In contrast, only indices of agility, but not

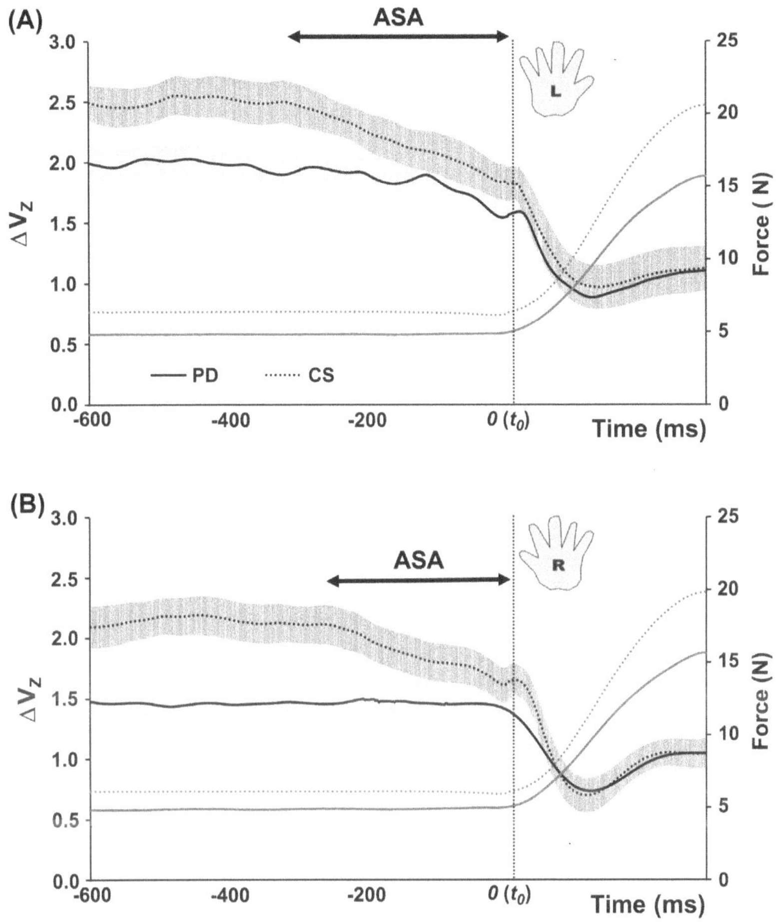

Figure 5.39 The time profiles of the index (ΔV_Z) of a multifinger synergy stabilizing total force in the left (A) and right (B, dominant) hands for a group of patients with early-stage Parkinson's disease (*solid line* with light standard error shade) and a group of age-matched control subjects (*dotted line* with dark standard error shade). Note the higher background level of ΔV_Z in the control group and longer and larger anticipatory synergy adjustments (ASA). Modified by permission from Park, J., Wu, Y.H., Lewis, M.M., Huang, X., Latash, ML., 2012. Changes in multi-finger interaction and coordination in Parkinson's disease. Journal of Neurophysiology 108 915–924. Copyright by American Physiological Society.

stability, improve with deep brain stimulation (DBS) of structures within the basal ganglia (Falaki et al., 2018). These observations suggest that DBS might lead to undesired consequences. Indeed, if patients with DBS become significantly more agile without a parallel improvement in action stability, they become more likely to lose stability in everyday tasks leading, in particular, to falls. Indeed, patients treated with DBS tend to show no clinically detectable improvement in postural stability and may fall more frequently (Weaver et al., 2009).

Similar problems with indices of stability and agility have been reported for patients with other neurological disorders involving loops through the basal ganglia and the cerebellum. In particular, smaller synergy indices and smaller ASAs have been seen in patients with multiple sclerosis and with multisystem atrophy (Park et al., 2013; Jo et al., 2017). Altogether, these observations suggest that unimpaired control of action stability depends crucially on functioning of brain loops involving the cerebellum, the basal ganglia, the thalamus, and different areas of the cortex. These loops have been described as crucial for functioning of so-called distributed processing modules within the scheme developed by Houk (2005).

Studies of patients after mild-to-moderate cortical stroke led to rather unexpected results, qualitatively different from those described for patients with subcortical disorders. The very first study (Reisman and Scholz, 2003) documented similar indices of synergies during reaching movements performed by the arm on the side of the lesion (ipsilesional) and by the arm on the opposite side (contralesional). This was particularly surprising because the movement patterns looked quite different on the two sides: The ipsilesional (less affected) arm showed reaching movements very similar to those in healthy control subjects, whereas the contralesional (more affected) arm showed significantly changed patterns with exaggerated involvement of the trunk. The lack of cortical stroke effects on indices of stability has been confirmed in a later study of multidigit tasks (Jo et al., 2016). That study, however, has demonstrated effects of stroke on indices of agility such as the duration and magnitude of ASAs (both were reduced in stroke survivors).

The main differences in the effects of subcortical (Parkinson's disease) and cortical (stroke) disorders on the indices of stability and agility are illustrated in Fig. 5.40.

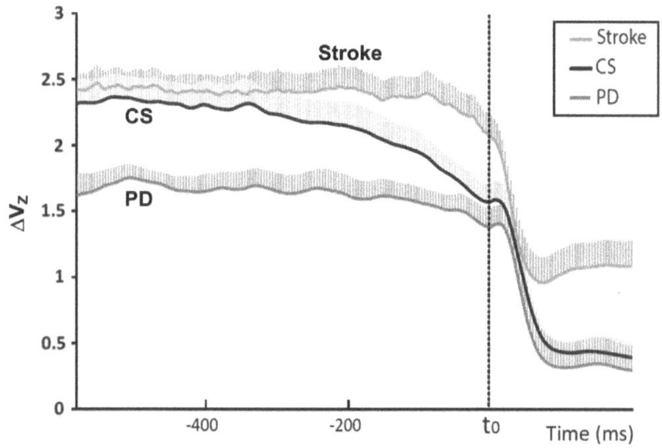

Figure 5.40 The time profiles of the index (ΔV_Z) of a multifinger synergy stabilizing total force in a group of patients with Parkinson's disease (PD), a group of patients after a mild cortical stroke, and a group of age-matched healthy control subjects (CS). Note the reduced ΔV_Z in the PD group but not in the stroke group. Anticipatory synergy adjustments (ASAs) are reduced in both PD and stroke groups as compared with the controls.

The subjects were asked to produce a constant level of total force while pressing with the four fingers of a hand and then, at a self-selected time, generate a very quick force pulse into a target. During steady-state force production, all subjects (healthy controls, those with Parkinson's disease, and those after stroke) stabilized total force by covarying commands to individual fingers as reflected by the relatively high values of the index of stability (ΔV). This index, however, was significantly lower in the Parkinson's disease group without a difference between the other two groups. Before the force pulse initiation, the control group showed a drop in the stability index starting 200–300 ms before the force pulse initiation time (shown as t_0), i.e., ASAs. The other two groups, however, showed significantly shorter and smaller ASAs.

Taken together, the reviewed clinical observations suggest that stability and agility may depend on different brain circuits. In particular, stability of action seems to depend crucially on subcortical loops. This conclusion is supported by observations of discoordination in patients with cerebellar and basal ganglia disorders leading to such signs as poor postural stability, ataxia, dystonia, chorea, etc. (reviewed in Fahn and Jankovic, 2007). Problems with agility, ability to initiate and switch movements quickly and effectively, seem to be equally inherent to cases of cortical and subcortical disorders.

So far, clinical studies of the ICS syndrome have been limited to analysis within multidimensional spaces of elemental performance variables, i.e., variables produced by apparent elements at the selected level of analysis. Analysis at the level of RCs described earlier (Section 5.6) has been limited to movements performed by young, healthy persons. Switching to analysis at the RC level could turn into a novel tool revealing the role of different brain structures in ensuring appropriate stability features of actions.

5.7.4 Using control tools in an unusual way

The abundant design of the human body allows performing various actions with different patterns of control variables leading to different involvement of elements such as limbs, joints, muscles, etc. Typical healthy persons show similar patterns during everyday actions, likely reflecting similarities of the anatomy, neurophysiology, and early childhood experience. However, even within this general population, there are substantial individual differences in movement patterns: Some people are more coordinated, whereas others are relatively clumsy. There are uniquely gifted athletes with movements that differ from those of an average control subject more than differences between typical healthy persons and those with mild neurological disorders. To summarize, "normal movement" is a misnomer; optimal movement patterns for any task depend on both person's state and external conditions of movement execution (reviewed in Latash and Anson, 1996).

It is natural, therefore, to expect neural control processes to differ among populations with atypical development, changed cognition, after highly specialized practice, etc. These differences can involve different RC patterns at the hierarchically highest task level and different few-to-many mappings involved in the generation of RC patterns at hierarchically lower levels (see Section 5.6). The former may be expected to result in atypical movement patterns evident in movement trajectories seen in averaged across trials performance. The latter may lead, in addition, to atypical patterns of action stability.

Consider a few features of atypical voluntary movements that are common across very different populations. These features include, in particular, slowness of movements and increased muscle cocontraction, which are commonly seen in healthy older person, persons with atypical development (such as Down syndrome), and patients with various neurological disorders. Note also that the same patterns are seen in young, healthy persons without specialized practice who find themselves in challenging conditions, for example, moving on ice or along a narrow path with precipice on both sides. This variety of populations suggests that the observed movement patterns are adaptive to different underlying conditions and represent attempts at avoiding complete failure at the expense of other factors, such as increased movement time and energy expenditure. Indeed, slowness of movements affords more time for planning and correcting movement patterns. Increased muscle cocontraction has two potentially helpful features: It improves stability of individual joints and extremities (although it may be detrimental for stability of the whole body, Yamagata et al., 2019a) and helps to organize action-stabilizing synergies at a control level (see Section 5.6.4).

Impaired control of movement stability described in the previous section is also common across numerous populations. In addition to healthy elderly and neurological patients, it is likely to be present in young adults with Down syndrome (Latash et al., 2002; Scholz et al., 2003). The likely involvement of subcortical loops, in particular, those via the cerebellum, in unimpaired control of stability makes it likely that populations with documented changes in the functioning of the cerebellum may also suffer from the ICS syndrome. These might involve persons not only with Down syndrome but also those with autism, Asperger's syndrome, developmental coordination disorder, and possibly other conditions affecting both cognition and motor coordination. ICS may also be expected in populations with likely involvement of cortical and whole-brain mechanisms, e.g., those with Alzheimer's disease and schizophrenia. Unfortunately, so far, information on possible changes in the control of action stability in those special populations is unavailable.

Kinesthetic perception

6

The traditional view on perception considers it a consequence of an interaction between two neural processes, afferent and efferent. The former term combines signals from sensory receptors, which may represent individual cells (as, for example, rods and cones in the retina and hair cells in the inner ear) or subcellular structures addressed as *sensory endings*. Receptors convert salient physical variables, e.g., those describing light, sound, displacements, and forces (and sometimes derivatives of those variables) into sequences of action potentials, which are then delivered into relevant structures within the central nervous system. Receptors can also respond to nonsalient variables. For example, they can be stimulated with electric current and with strong mechanical deformation: A strong hit of the eye is commonly associated with a visual perception of flying stars or sparkles.

In particular, kinesthetic perception is based on signals from proprioceptors, specialized neurons located in the spinal ganglia (for the trunk and extremities), which have a long T-shaped branch termed *axon* (Fig. 6.1). The peripheral part of the axon ends with a sensory ending sensitive to deformation and sometimes to other factors (e.g., inflammation, products of muscle metabolism, temperature, etc.). Specificity of sensory endings to different salient mechanical variables is defined by three main factors. The first factor is location of the sensory ending in the body. The second is relative sensitivity of the ending to deformation rate. And the third is adaptation, a phenomenon reflecting a reduction of the ending's response to an ongoing stimulus with time.

There has been little disagreement on the role of afferent signals for perception, although their specificity for particular mechanical variables such as joint angle, muscle length, muscle force, etc., has been debated. Indeed, several recent studies have suggested that sensory endings traditionally viewed as receptors for muscle force (Golgi tendon organs) can serve as length receptors, and vice versa, receptors located in muscle spindles traditionally viewed as those of muscle length and velocity can serve as force receptors (Watson et al., 1984; Luu et al., 2011; Proske and Gandevia, 2012; Rescshechtko et al., 2018). Besides, a number of studies have documented convergence of signals from different afferent systems on common spinal interneurons (Jankowska, 1979; Lundberg 1979; Jankowska and McCrea, 1983; Kistemaker et al., 2013), which would seem to garble information from sensory endings sensitive to different mechanical variables.

In contrast, the role of efferent processes in kinesthetic perception has been more controversial. There are numerous examples of effects of action of perception; some of them are discussed later. However, there is no agreement in the research community on the organization of action-related neural processes and even on what neural variables are essential in producing a voluntary movement. In previous parts,

Physics of Biological Action and Perception. https://doi.org/10.1016/B978-0-12-819284-9.00006-1

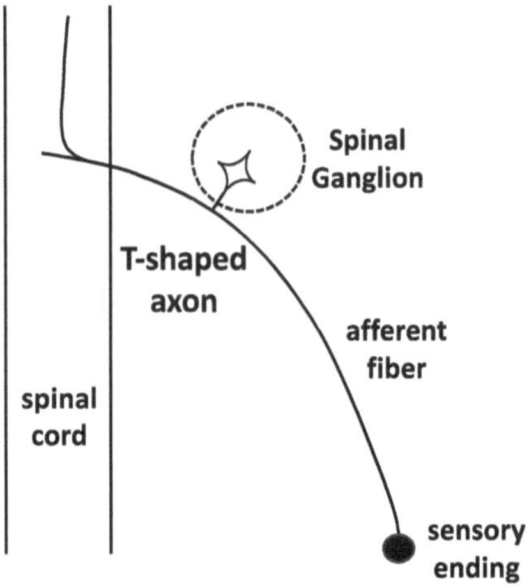

Figure 6.1 A proprioceptor neuron has the body located in the spinal ganglia and a T-shaped axon with two long branches. The peripheral branch ends with a sensory ending, whereas the central branch enters the central nervous system with effects at both spinal and brain levels.

a particular view on the neural control of voluntary movements has been presented, and, in this part, I will follow the same philosophical approach and try to generalize it to kinesthetic perception.

6.1 Main proprioceptors and what they measure

There are several groups of sensory endings that generate signals with mechanical variables more or less directly related to body configuration and contact forces exerted on the environment. Such receptors are located in the skin, subcutaneous tissues, joint capsules, muscles, and muscle–tendon junctions. While their location suggests relevance to specific mechanical variables, conclusions based on location only are sometimes misleading. Arguably, the most obvious example is the joint capsule receptors.

The location of joint capsule receptors makes them perfect candidates for reporting joint angle, which is one of the crucial variables for perception of body configuration. Projections of the central axon branches of the corresponding proprioceptor neurons to the sensorimotor cortex were discovered relatively early and were used as an additional argument that joint capsule receptors were the main source of joint position sense.

This view took the first strong hit in the 1970s when direct recording from the axons innervating joint receptors showed that most of these receptors changed their frequency of firing when joint angle approached limits of its rotation, not in the midrange of joint position (Burgess and Clark, 1969; Clark and Burgess, 1975). In addition, these receptors were highly sensitive to joint capsule tension and joint inflammation. So, when a person co-contracts agonist–antagonist muscles without moving the joint, joint receptors are expected to show an increase in their firing rate while the joint remains motionless. If, indeed, joint position senses were based on signals from those receptors, how could the central nervous system disambiguate the effects of joint rotation and joint capsule tension? Any inflammatory process involving a joint should also affect joint position sense, which is not typically observed.

More evidence against the role of joint receptors in joint position sense came from observations in persons after total joint replacement: These persons showed no visible deterioration in joint position sense, despite the fact that all the receptors housed in the joint were removed from the body (Cross and McCloskey, 1973; Grigg et al., 1973). Currently, the dominant view is that, despite their seemingly perfect location, joint receptors contribute little to joint position sense. They are more likely to play the role of safety sensors informing the central nervous system on situations that potentially can lead to joint damage such as those associated with being close to the anatomical limit of joint rotation, experiencing high joint capsule tension, and being affected by joint inflammation.

According to most recent books and review articles, crucial afferent information for joint position sense is believed to be supplied by sensory endings located in muscle spindles, small structures scattered all over skeletal muscles (Fig. 6.2). Spindles represent structures with a shell made of connective tissue; the shell is connected to power-producing muscle fibers that are referred to as *extrafusal* fibers, i.e., those located outside the spindle. The intrinsic design of muscle spindles is rather complex. Spindles contain muscle fibers, addressed as *intrafusal* fibers, with sensory endings sensitive to deformation located on those fibers. The intrafusal fibers connect to the inside of the spindle shell and do not contribute to tendon force and power production. When extrafusal fibers change their length (this happens when the muscle is stretched or shortened), the spindles experience deformation, which is transmitted to the

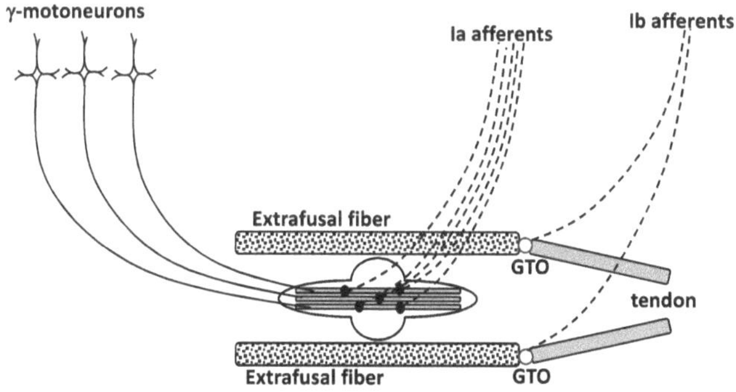

Figure 6.2 Muscle spindles are oriented parallel to the power producing extrafusal muscle fibers. The spindles contain sensory endings innervated by fast-conducting Ia afferents and by group II afferents. The endings are located on intrafusal muscle fibers, which are innervated by γ-motoneurons. Golgi tendon organs are located in the muscle–tendon junction, in series with the extrafusal fibers. They are sensitive to force and innervated by group Ib afferent fibers.

intrafusal fibers and sensory endings. The sensory endings change their frequency of generation of action potentials in response to the deformation.

There are two types of sensory endings in muscle spindles. The so-called primary endings are sensitive to both muscle length and velocity, while the secondary endings are sensitive to length only. Note that sensitivity to velocity implies that the endings increase their activation level when the muscle is stretched and decrease the activation level when the muscle is shortened. In particular, during quick muscle shortening, primary endings show a transient period of silence. The primary endings are innervated by very thick axons, known as *Ia afferents*, which conduct action potentials at a very high speed, up to 120 m/s; this makes Ia afferents the fastest conducting neural fibers in the body. The intrafusal muscle fibers are innervated by small neurons in the spinal cord, so-called *gamma-motoneurons*, which produce contractions of those fibers (see Fig. 6.2). When gamma-motoneurons induce contraction of intrafusal fibers, the fibers change their mechanical properties and their response to deformation under external forces. As a result, the system of gamma-motoneurons allows controlling sensitivity of the spindle endings to muscle length and velocity.

The sophistication of the spindle design suggests that proper signals from spindle endings are crucial for some of the most important functions of the body. The fact that some of those endings are innervated by the fastest conducting neural fibers strengthens the conclusion that these signals are not only vital but also urgent.

Another group of sensory endings innervated by fast conducting fibers, *Ib afferents* (Fig. 6.2), are located at the junction between the muscle fibers and tendon. These endings, called *Golgi tendon organs*, are also sensitive to deformation. However, while spindles are located in parallel to the muscle fibers, Golgi tendon organs are located in series to those fibers. This location makes them sensitive to tendon force. Indeed, Golgi tendon organs change their firing frequency with force with no sensitivity to force rate. These receptors do not have a system that would be analogous to gamma-motoneurons; so, their sensitivity to force is not modulated by the central nervous system.

Several types of receptors are located in the skin and subcutaneous tissues. These receptors are sensitive to deformation and differ from each other by their low or high sensitivity to the rate of skin deformation, sensitivity to normal and tangential to the skin directions of deformation, large or small areas of sensitivity, and quick or slow adaptation. For example, the Puccini corpuscles are large, highly sensitive to quickly changing skin deformation (for example, during vibration), do not have high spatial resolution, and adapt quickly (meaning that their response to ongoing deformation quickly disappears). On the other end of the spectrum are the Merkel disks, which are small, sensitive to vertical pressure on the skin, and slowly adapting, which means that they can supply continuous information on pressure on the skin for long periods of time.

Overall, it seems that the bodies are equipped with a large number of types of sensory endings that are sufficient to provide information on all salient mechanical variables. This impression, however, looks misleading if one considers salient variables for joint movements, which are typically rotations. These salient variables are joint angle (and its derivatives) and moment of force. Note that there are no sensory endings that would provide accurate information on those variables. For example, Golgi tendon organs provide information on tendon force, but moment of force depends also on the lever arm of the force vector. The lever arm changes with joint angle. As a result, the same activity level of Golgi tendon organs can correspond to different moments of force if observed at different joint angles.

Along similar lines, muscle spindles may be seen as providing information on muscle fiber length. This information, however, is by itself ambiguous given the availability of gamma-motoneurons. Note that muscle activation is typically accompanied by an increase in the activity level of gamma-motoneurons (alpha-gamma coactivation). So, if a person activates a muscle, gamma-motoneurons increase their activation level and change afferent signals from muscle spindles even if no motion takes place. Note also that joint angle is coupled to the length of the "muscle + tendon" complex, and muscle fiber length can change even if the length of this complex remains constant. For example, during a strong contraction in isometric conditions (no joint motion), muscle fibers shorten while the tendon is stretched. Relative changes in the length of the tendon and muscle fibers during movement depend on the level of muscle force: Muscle fibers are typically more compliant when the muscle is relaxed and stiffer when the muscle is activated.

All the mentioned factors suggest that afferent information, by itself, supplies messy and ambiguous information relevant to salient variables for kinesthetic perception. On the other hand, without this information, both perception and action show major impairments. This was known for a long time based on the aforementioned observations of patients with *tabes dorsalis*, a consequence of untreated syphilis common in the first half of the 20th century. These patients lost transmission along the dorsal columns of the spinal cord that carried sensory information into the brain. They showed major impairments of functions that require sensory information, such as standing and walking with one's eyes closed.

Another mentioned disorder of the peripheral sensory system (large-fiber peripheral neuropathy) is a rare condition associated with loss of signal conduction along thick peripheral neural fibers that carry information from many types of sensory endings. Such *"deafferented patients"* cannot feel touch to their limbs and limb motion, while muscle force is not significantly affected. These patients can perform functional movements with the help of vision, but their movements are typically slow and awkward. Meaningful movements become impossible when visual information is unavailable. Taken together, these observations demonstrate the crucial importance of information from peripheral receptors not only for perceptual processes but also for the control of movements.

6.1.1 Problems with measuring percepts

One of the big problems in developing a framework for understanding perception based on laws of nature is the inherent difficulty of measuring percepts. Indeed, in studies of movements, it is possible to measure many salient variables such as coordinates and forces with reasonably high accuracy. Other variables, not directly measurable (e.g., muscle forces), can be computed from the measured ones based on mechanical models and estimated salient parameters. In studies of perception, it is possible, although by no means easy, to measure physiological signals contributing to perception at different levels, from sensory endings to brain structures. However, as we will see further in this section, there seems to be no one-to-one correlation between any of those signals and what humans report as perceived variables. It is also possible to measure some of the perceived variables objectively, for example, mechanical variables (e.g., weight, force, distance, etc.). However, perception is by definition subjective: The same object may feel "light" to one person and "heavy" to another person or even to the first person in a different state, e.g., under fatigue. So, perception has remained an axiomatic notion without an exact definition. We are going to consider a definition later within this part of the book based on a particular approach to the nature of perception.

There have been two major approaches to measuring percepts. The first one is based on developing a psychophysical scale, i.e., a subjective scale, which is used by the person to report the magnitude of a percept. Such scales have been developed to report percepts of physical variables, for example, weight and size of objects held by the hand, spatial position and orientation of body parts, loudness of a sound, luminance of a source of light, etc. (reviewed in Borg 1982; Geschelder, 1997; Krumhansl, 2000). They have also been used to report intrinsic variables such as perception of pain. In most studies, psychophysical scales have been expressed not in objective physical units, such as kilograms and meters, but in units within artificial scales (e.g., a 10-point or a 100-point scale) expressed relative to the person's abilities, body size, strength, and other objective or subjective measures. For example, psychophysical scales for pain have commonly been developed with respect to the maximal level of pain one can endure. In different studies, participants reported the magnitudes of percepts by speaking out a number on the corresponding scale or using other means, e.g., stopping a signal on the computer screen when it reached the perceived value.

Developing a psychophysical scale typically requires explanation and training, but even after extensive practice, scales are not very reproducible and rarely linear. They drift with time and depend on the recent history of person's activities including exposure to signals of the same modality. For example, typical daylight may seem extremely bright and lead to, respectively, high psychophysical scores if one comes out of a very dark cave.

The alternative to psychophysical scales is comparing percepts to a set of standard stimuli or reporting percepts by one's own actions. For example, if one wants to measure accuracy of perception of force applied by an effector to an external object (e.g., pressing with a finger), the participant may be asked to relax and then produce the same force again. This is, of course, not a very reliable method because it involves memory. Another option is to ask the person to match the force applied by a finger by pressing with the symmetrical finger of the other hand. While this method avoids using memory, it involves another potentially confounding factor such as the difference between the two hands (or any two effectors). In addition, participants in such experiments may also reinterpret the instruction as matching "commands" to the two effectors rather than percepts (e.g., van Doren, 1998), which obviously can lead to misleading conclusions if one is interested in estimating percepts.

The lack of a reliable measuring device for salient variables is not unique for studies of perception. Indeed, a major problem with the equilibrium-point hypothesis, which may be blamed for preventing it from being broadly accepted by the research community, has been the lack of "lambda-meters," i.e., objective and reliable tools to measure salient variables. The same comment applies to the general theory of control with referent coordinates (RCs): RCs are typically not directly observable, only their indirect effects on movement mechanics are. This has been a challenge in movement studies, but a number of approaches have been developed to get indirect estimates for such hidden variables as λs and RCs (Latash and Gottlieb, 1990; Latash et al., 1999; Ambike et al., 2015b, 2016a). It may be a matter of time before reliable measures of percepts are developed; meanwhile, we have to rely on the described methods such as those involving matching actions and using psychophysical scales.

6.2 Efferent copy

One of the most influential concepts relevant to the role of efferent processes in perception has been that of the *efferent copy* (von Holst and Mittelstaedt, 1950), sometimes also addressed as *efference copy* or *corollary discharge* (Sperry, 1950). The original notion was based on separating afferent signals into two categories, *exafference* and *reafference* (Fig. 6.3). Exafference (ExA in Fig. 6.3) was assumed to reflect changes in the muscle state because of changes in external forces in the absence of an

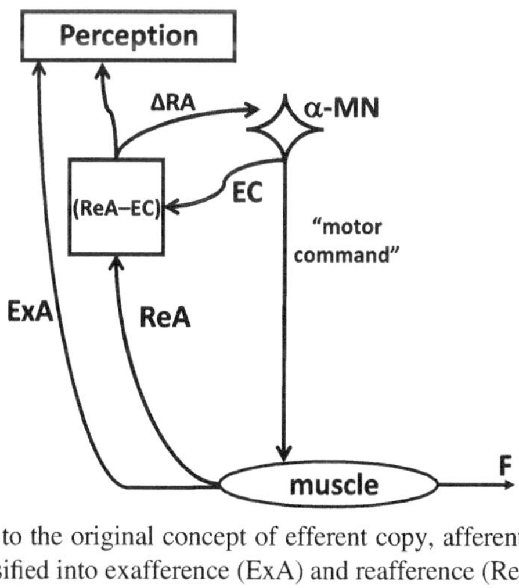

Figure 6.3 According to the original concept of efferent copy, afferent signals from proprioceptors are classified into exafference (ExA) and reafference (ReA). Exafference reflects movements produced by external forces (F), while reafference reflects effects of voluntary muscle activation ("motor command", output of alpha-motoneurons, α-MN). A copy of the "motor command" (efferent copy, EC) is used to predict changes in afferent information. The difference between ReA and EC produces an error signal, which contributes to perception and to changes in the "motor command."

intentional action, while reafference (ReA in Fig. 6.3) reflected effects of self-initiated actions on sensory signals. Within this scheme, exafference always affects perception, while reafference has effects on perception only if it deviates significantly from changes in the afferent signals expected from the self-generated, ongoing motor command. The term *efferent copy* was used to address a hypothetical mechanism underlying such expectations.

More recently, a similar idea has been developed in the form of internal direct models: hypothetical mechanisms that predict changes in body state and associated sensory signals based on the ongoing efferent process (Imamizu et al., 1998; Wolpert et al., 1998; Kawato, 1999). While the concept of internal models remains purely theoretical—they assume computations within the central nervous system with no clear physiological substrate—the original concept of the efferent copy associated it with a copy of the current motor command to a muscle, i.e., with the output signals from alpha-motoneurons. According to the original scheme, these signals were used by the central nervous system to estimate sensory signals during the action, i.e., reafference. The estimation process was assumed to participate in both perceiving the muscle state and correcting the action if there was a discrepancy between the expected and actual changes in sensory signals. As such, the efferent copy was assumed to play an important role both in perceptual processes and in modifications of commands to muscles in response to changes in sensory information.

Although the general idea of estimating afferent signals with respect to the current efferent signals remains valid, the original scheme has been criticized recently (Feldman, 2009, 2015, 2016). In particular, the original scheme does not allow relaxing muscles after a movement to a new position. Indeed, imagine that a person performs a joint motion from a relaxed state and then relaxes in the new joint position. As both muscles acting at a joint are relaxed in both positions, the efferent copy (a copy of the signals from alpha motoneurons) is the same (zero!) in both positions, while muscle length values are different. Sensory signals associated with muscle length, such as those from muscle spindle endings, should generate different activity levels. If these signals, compared to the efferent copy, produced no error signals in the initial joint position, they have to produce a non-zero error signal in the final position leading to non-zero muscle activation. This prediction does not allow relaxing in different joint positions in clear conflict with both everyday experience and experimental observations.

One of the first persons to appreciate the nontrivial ability of humans to relax muscles in different joint positions was Kurt Wachholder who performed (with his student, Altenbürger) an experiment in 1920s and showed that humans were indeed able to relax muscles in different postures (Sternad, 2001). At that time, muscle elasticity had been already known, and Wachholder concluded that the central nervous system had to be able to modify muscle spring-like properties during voluntary movement—a prediction making that old study one of the direct predecessors of the equilibrium-point hypothesis.

An alternative interpretation of interactions between afferent and efferent processes for kinesthetic perception has been developed within the framework of the equilibrium-point hypothesis and the idea of movement control with RCs (Feldman and

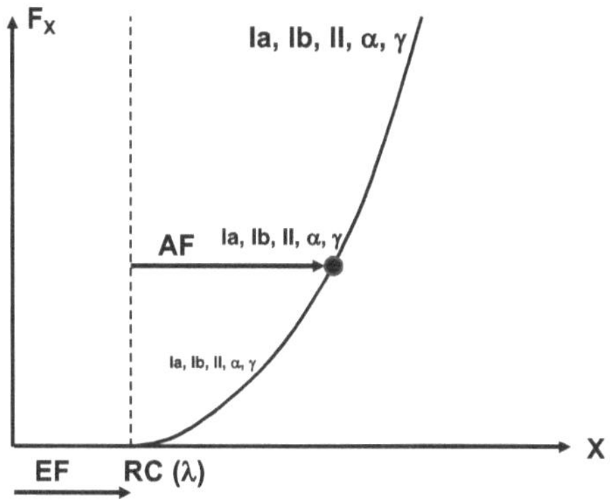

Figure 6.4 Kinesthetic perception is based on two components, efferent (EF) and afferent (AF). The former defines a referent coordinate (RC), while the latter defines deviations from RC along the force–coordinate relationship. For a single muscle, RC is equivalent to threshold of the tonic stretch reflex, λ. Many neural signals change monotonically along the force–coordinate, $F_X(X)$ line; these include both afferent and efferent signals.

Latash, 1982a,b; Feldman, 2009). Fig. 6.4 illustrates this concept for an arbitrary spatial vector variable $\boldsymbol{X}$ and associated force vector $\boldsymbol{F_X}$. The efferent process (EF in Fig. 6.4) sets a value of RC, which may be viewed as a frame of reference within which afferent signals (AF in Fig. 6.4) are quantified. This scheme is a particular realization of the basic idea that to measure anything you have to have a system of coordinates and a measuring device (e.g., a referent point and a ruler). It is compatible with the idea of efferent copy but associates it not with the output of alpha-motoneuronal pools but with RC, which is equivalent to threshold of stretch reflex (λ) at the single-muscle level.

The scheme illustrated in Fig. 6.4 at the level of a single muscle assumes that the senses of muscle position and force are tightly coupled and emerge as a result of interaction between the same two processes, efferent and afferent. The efferent process specifies a RC (λ), while the afferent process reflects deviations of the system from the RC. Indeed, setting a value of λ directly contributes to solving the problem of perceiving muscle length and force: For a given λ, only points on the corresponding $F(L)$ muscle characteristic are possible as equilibrium states of the system "muscle + reflexes + load." Effectively, setting a value of λ reduces the theoretically possible two-dimensional array of force and length values to a single-dimensional subspace—the $F(L)$ curve.

All the major sources of sensory information are expected to increase with an increase in the distance of points on the $F(L)$ curve from λ. In particular, moving along the $F(L)$ curve from λ to the right in Fig. 6.4 is associated with an increase in muscle length, which is expected to lead to an increase in the activity level of muscle spindle sensory endings (shown in Fig. 6.4 as an increase in activity of Ia and II afferents). At the same time, muscle force increases leading to an increase in the activity of Golgi tendon organs innervated by Ib afferents. Muscle activation level also increases along the $F(L)$ curve accompanied by higher activation levels of both alpha-motoneurons and gamma-motoneurons (because of the aforementioned phenomenon of alpha-gamma coactivation).

So, any of the aforementioned signals can be used to identify a point on the $F(L)$ curve and result in accurate perception of both muscle length and force. In other words, although specific sensory endings are sensitive to only some of the salient variables, estimating them within the reference frame set by λ allows perceiving mechanical variables that have no direct effect on the activation level of those sensory endings. Indeed, as it follows from Fig. 6.4, spindle endings may be seen as force sensors, Golgi tendon organs—as muscle length sensors, and even activity from the output neurons of the efferent process, alpha- and gamma motoneurons, can be used for perceiving F and L.

This approach to kinesthetic perception makes the space of sensory signals contributing to kinesthetic perception abundant. As in any abundant system (see the discussion on abundance in Part 3), this does not lead to computational problems such as ambiguity in interpreting each signal individually. In contrast, abundance makes perceiving a point on the $F(L)$ curve robust to possible problems with any one of the signals, for example, because of a pathology or surgery. This view is supported by the aforementioned observations of nearly unchanged joint position sense following total joint replacement surgery (Cross and McCloskey, 1973; Karanjia and Fergusson, 1983; Wada et al., 2002).

One of the nontrivial predictions of the scheme illustrated in Fig. 6.4 is that illusions associated with artificial changes in the outputs of particular groups of sensory endings should have both spatial and force components, even when the affected sensory endings are directly sensitive to one group of these mechanical variables only. This prediction has been confirmed in studies of vibration-induced kinesthetic illusions (VIKI) described in the next section.

6.3 Vibration-induced kinesthetic illusions

High-frequency, low-amplitude vibration has been known for a long time as a powerful stimulus for primary spindle endings because of their sensitivity to muscle velocity (Brown et al., 1967; Matthews and Stein, 1969). Vibration at about 100 Hz with the peak-to-peak amplitude of 1 mm applied to the belly of a muscle or to its tendon can drive all the primary spindle endings at the frequency of vibration and sometimes even induce a few action potentials per vibration cycle. The unusually high activity level of the primary spindle endings brings about a number of motor and perceptual consequences. The former includes the tonic vibration reflex (TVR) seen most commonly in the muscle group subjected to vibration (Eklund and Hagbarth, 1966). Reflex muscle activation can also be seen in other muscles of the limb depending on the initial limb configuration and a few other factors (Latash and Gurfinkel, 1976). These patterns have suggested an involvement of spinal neural circuits involved in the generation of locomotor movements, an idea supported by the observations of locomotor-like leg movements in humans during vibration of one of the major leg muscle groups (Gurfinkel et al., 1998; Selionov et al., 2009; Solopova et al., 2016).

Perceptual effects of muscle vibration are also many and varied. Most commonly, kinesthetic illusions are reported corresponding to perceiving joint positions associated with longer length of the muscle subjected to vibration (Goodwin et al., 1972; Roll and Vedel, 1982; Lackner and Taublieb, 1984). Such illusions can be very strong and even lead to perception of anatomically impossible joint configurations (Craske, 1977). Interpretations of such VIKI typically invoke the unusually high level of primary spindle endings, which are sensitive to both muscle length and velocity. Velocity-sensitivity probably plays a particular important role. Indeed, in some studies, humans reported very high velocities of illusory movements produced by muscle vibration, whereas illusions of joint displacement were relatively modest, much smaller than the product of the reported joint velocity and time (Sittig et al., 1985).

If vibration is applied to a muscle involved in a postural task, for example, to the Achilles tendon during standing, humans feel that the body is pulled backward and show significant body deviations from the vertical, which sometimes require making a step (Eklund and Hagbarth, 1967; Hayashi et al., 1981). Such vibration-induced fallings (VIFs) have also been interpreted as consequences of perceptual effects of vibration. Indeed, during vibration of the Achilles tendon, the brain is expected to associate the unusually high activity levels of the primary spindle endings in the ankle plantar flexors with elongation of that muscle group, which is associated with body sway forward. This illusory perception is corrected by an actually observed deviation of the body in the opposite direction, i.e., VIF.

A number of features of VIKI have remained unexplained. For example, the direction of VIKI could reverse with changes in sensory signals of other modalities and even with a cognitive task performed during muscle vibration (Feldman and Latash, 1982c). Besides, humans vary broadly in their sensitivity to muscle vibration: Some people report very strong VIKI, whereas others feel only mild illusions. Some VIKI-related illusions are seen consistently in only a small proportion of the general populations. These include, in particular, the so-called Pinocchio effect (Burrack and Brugger, 2005). This illusion is sometimes reported if a person is asked to touch lightly the tip of the nose with the index finger of a hand and then vibration if applied to the elbow flexor group (e.g., over the biceps). The high activity level of the spindle endings in the elbow flexors suggests to the subject that the elbow is being extended. On the other hand, the contact between the fingertip and the nose is not broken. The combination of the two factors leads, in some persons, to reporting an illusion of nose elongation.

Only some of the perceptual effects of vibration allow relatively easy interpretation based on the increased activation of primary spindle endings creating an illusion of the muscle being stretched and leading to reports of corresponding joint rotations and other effects. Other effects suggest an important role of vibration effects on the efferent component of perception, i.e., on RC (cf. Fig. 6.4; Feldman and Latash, 1982a; Reschechtko et al., 2018).

Indeed, consider the fact that any joint is crossed by multiple muscles. Although vibration spreads through soft tissues, its effectiveness drops rather dramatically with distance from the vibration site. So, in a first approximation, we can assume that sensory receptors in muscle groups that are not directly subjected to vibration

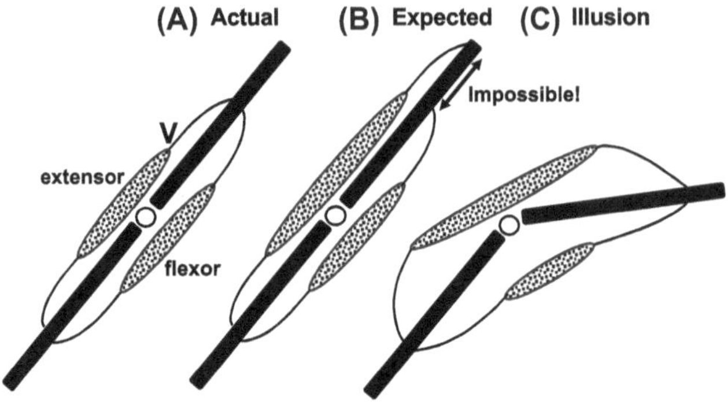

Figure 6.5 If vibration (V) is applied to a muscle (panel A), its length is perceived as increased. Perception of the length of the antagonist muscle is assumed to be unaffected. This should lead to perception of an anatomically impossible change in the muscle attachment point (panel B). Actual illusion corresponds to changes in the length of both muscles (panel C).

are providing veridical information about the actual muscle length and joint position. For example, if a vibrator is placed over the elbow extensors (Fig. 6.5A), sensory information from spindle endings in the extensor muscles may be interpreted as reporting that those muscles are being stretched at a high velocity. At the same time, afferent signals from the spindle sensory endings in the flexors suggest that the length of the flexor muscle group does not change (see Fig. 6.5B). This is an impossible combination of sensory signals, which violates joint integrity: Taken at face value, it suggests that some of the muscles detached from their natural insertion points and/or that bone deformation happened. Such illusions are not reported, however. Rather, the seemingly impossible combination of afferent signals leads most commonly to a consistent illusion of joint rotation corresponding to stretch of flexors and shortening of extensors (Fig. 6.5C). These observations lead to several conclusions.

First, it seems that the neural mechanisms responsible for perception avoid percepts of deformed or disjointed body even if changes in afferent information suggest that this is happening. Interestingly, there are exceptions from this rule. For example, anatomically impossible rotations are not precluded: As mentioned earlier, sometimes humans report perceiving impossible joint rotations such as elbow hyperextension (Craske, 1977; besides, the author of this book had firsthand experience with perceiving impossible shoulder positions under vibration of the upper arm muscles). Another exception is the aforementioned Pinocchio effect suggesting an impossible change in the body anatomy.

Second, the large variability of VIKI across people and the possibility of VIKI reversal suggest that VIKI are inherently unstable. This conclusion is also supported by the observations of no consistent illusion with an increase in the intertrial variability of position sense when vibration is applied simultaneously to agonist and antagonist muscle groups (Fuentes et al., 2012). This leads us to the next sections, which try to develop the concept of stability for multielement systems to perceptual processes.

6.4 Perceptual stability

Stability of perceiving the external world and one's own body in that world is clearly very important for a variety of functions. Otherwise, purposeful movements in a wobbly world, for example, reaching for a target, would be impossible or, at the very least, very complicated and requiring multiple corrections. However, accurate movements to a target are seen across species and tasks, even in preparations with the spinal cord surgically separated from the brain (spinal preparation). For example, the spinal frog can produce an accurate wiping movement by a hindlimb directed at the irritating stimulus placed on its skin (Berkinblit et al., 1986). Accurate wiping is also observed when the stimulus is placed on a forelimb, and the position of the forelimb with respect to the trunk is changed from trial to trial (Fukson et al., 1980). These observations suggest that even the spinal cord is able to create an accurate reflection of the body configuration and to use this reflection to direct motor actions. What could be the origin of such veridical and stable percepts?

Imagine that you press with an effector, e.g., a hand, against a stop (e.g., as in Fig. 6.6). While you vary the pressing force, you have a stable and veridical perception of all the joints being motionless (of course, the eyes are closed in this mental experiment.) Note that all the sensory (afferent) and motor (efferent) signals are changing with force in this experiment. Indeed, the relative length of muscle fibers changes (they shorten in contracting muscles while the tendon stretches, compare the black and gray muscle shapes in Fig. 6.6), which is expected to produce changes in the signals from the spindle sensory endings. Besides, gamma-motoneurons change their activation levels in parallel to alpha-motoneurons (alpha-gamma coactivation), which is also expected to produce changes in the spindle afferent activity reflecting muscle length and its changes. Golgi tendon organs change their activity levels with muscle force; note that pressing against a stop is associated with non-zero levels of agonist–antagonist coactivation, thus leading to changes in force-related afferent activity from both muscle groups (not shown for the antagonist muscle in Fig 6.6

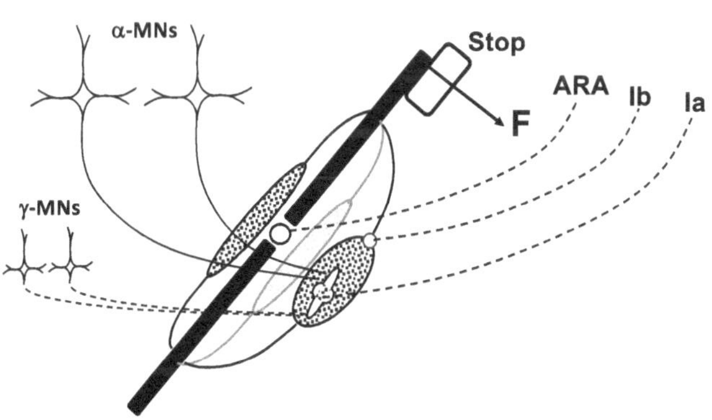

Figure 6.6 When an effector presses against a stop with force F, signals from all proprioceptors, from spindles (Ia), Golgi tendon organs (Ib), and articular afferents (ARA) change as well as efferent signals from both alpha- and gamma-motoneurons α-MNs and γ-MNs.

for simplicity). Articular receptors are sensitive to joint capsule tension, which is also expected to change during this mental experiment. Efferent signals also change. Alpha-motoneurons produce muscle activation, while gamma-motoneurons show parallel changes in their activity. It seems, therefore, that there is no signal within the body that would be reliably constant during pressing against a stop. So, how do we know that the elbow does not move?

There is no direct answer to this question, but the observation of stable perception of joint position in such conditions (and also during voluntary cocontraction of the agonist–antagonist muscle pair crossing the joint) suggests a particular organization of all the relevant signals, afferent and efferent. Moreover, the idea of hierarchical control of actions with RCs for the effectors suggests a related hierarchical scheme of perception.

6.5 Transformations involved in actions and perceptions

Processes involved in both action and perception are associated with transformations among systems of coordinates. For example, the sensory endings in the muscles, tendons, joints, and skin are typically sensitive to local events, such as the deformation of tissues adjacent to the sensory ending, and produce sequences of action potentials reflecting mechanical variables, such as local forces and displacements. This step may be viewed as transformation between a system of coordinates reflecting mechanical variables (forces and coordinates) and a system of coordinates expressed in neural variables (sequences of action potentials). Furthermore, signals from sensory ending, likely in combination with efferent signals, produce percepts reflecting the body configuration and its interaction with the environment, which may be expressed in another set of salient mechanical variables, forces, and coordinates. In summary, kinesthetic perception involves a sequence of transformations: mechanical (actual) => neural (electrophysiological) => mechanical (perceived).

With respect to action, some of the transformations involved in the neural process have been discussed in earlier sections. For example, the neural control process has been viewed as starting with specifying a task-related trajectory of the RCs for a few salient variables, $RC_{TASK}(t)$. These may be related to a desired body trajectory in space, as in locomotion, or to trajectory of the hand as in reaching toward a target, or to forces produced on the environment as in gripping a handheld object. Further $RC_{TASK}(t)$ leads to trajectories of limb-level, digit-level, and joint-level control variables, such as $r_j(t)$ and $c_j(t)$ at the joint level. Ultimately, those have to be converted into muscle-level variables, $\lambda_i(t)$. These neural variables are converted into movement with important roles played by the stretch reflex and the external force field.

Consider a simple everyday action, for example, picking a glass with wine or another favorite liquid. This action involves multiple steps including those that rely on adequate perception of one's own body parts in the environment. For example, if one of the hands is involved in another task, the other hand has to be selected as the effector for the new task. Furthermore, the glass with wine has to be selected from potentially numerous objects in the visual field. The actor has to define when to initiate the action and its speed, generate a movement toward the glass, and introduce corrections in cases of movement errors or changes of the target object location (for example, if the glass stands on the tray carried by a waiter). While we know very little about the exact neural processes involved in all these steps, it is obvious that multiple few-to-many and many-to-few transformations are involved in such an action and associated perceptual processes.

Let us start with the lowest, somewhat arbitrarily selected, functional level, i.e., that of a single muscle (Fig. 6.7). The number of salient physical variables describing muscle state (such as length, force, and maybe their derivatives) is relatively small. It is much smaller than the number of afferent and efferent signals, generated by individual neurons, which contribute to perception of those physical variables (hundreds or even thousands). So, this step involves a many-to-few transformation: $\{AF; \lambda\}_i \Rightarrow \{PV_M\}_j$, where the subscripts i and j indicate dimensionality of the sets of variables ($i \gg j$; note that in Fig. 6.7, $i = i1 + i2$), AF stands for all the relevant afferent modalities (e.g., signals from muscle spindles, Golgi tendon organs, and other muscle-level sensory endings), and PV_M stands for salient perceived variables at the muscle level.

Kinesthetic perception at the joint level also can be seen as involving a relatively small number of salient physical variables (e.g., joint angle, velocity, and moment of force). At this level, the number of afferent sources of information becomes even larger because of the articular receptors and the fact that typical joints are spanned by several muscles with the corresponding sets of spindles and Golgi tendon organs. At the control level, a simple joint with a single rotational degree-of-freedom is typically served by more than two muscles, so the number of relevant λs is larger than two. An alternative pair of neural commands have been introduced, the reciprocal and coactivation commands (*r*- and *c*-commands), which produce changes in the joint referent angle and its apparent stiffness, assuming $c > 0$ (Feldman, 1980). At this level, the many-to-few transformation can be expressed as $\{AF; r,c\}_n \Rightarrow \{PV_J\}_m$, $n \gg m$; PV_J stands for salient perceived variables at the joint level (in Fig. 6.7, $n = n1 + n2$).

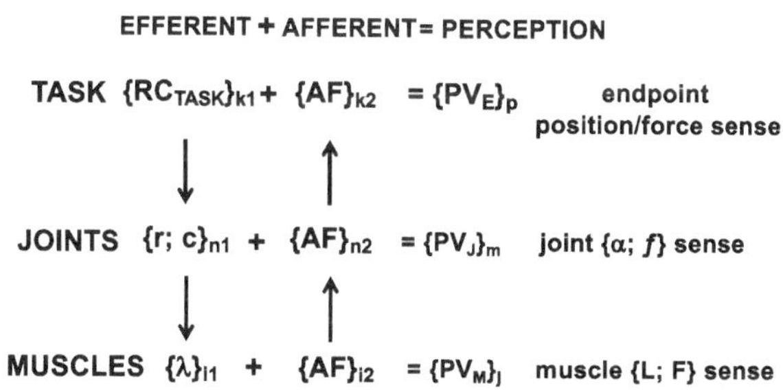

Figure 6.7 A schematic illustration of perception at different levels, from single muscle (bottom), to joint (middle), and to a multijoint task-relevant effector (top). At each level, the number of elemental variables, afferent plus efferent, is larger than the number of salient variables relevant to perception.

The abundance of the afferent sources at the joint level leads to the very mild, if any, effects of surgical total joint replacement on joint position sense (Cross and McCloskey 1973; Karanja and Fergusson, 1983). This procedure leads to the elimination of many sources of afferent information relevant to joint position sense including receptors located in the joint capsule and in other tissues removed during the surgery, but the remaining sources are obviously sufficient to ensure nearly unchanged perception of joint position.

At the limb level, the number of relevant afferent signals continues to grow, whereas the number of salient mechanical variables remains relatively low: $\{AF; RC\}_k => \{PV_E\}_p$, $k \gg p$; PV_E stands for salient performance variables at the effector level (in Fig. 6.7, $k = k1 + k2$). So, at each of the levels, the number of signals, both afferent and efferent, contributing to perception of salient physical variables is always much larger than the number of those physical variables. This means that, at each level, the central nervous system handles abundant sets of elemental variables.

6.5.1 The Kohnstamm phenomenon and podokinetic effect

The entanglement of action and perception processes is potentially a major contributor to a number of phenomena in the field of unintended movements. One of them is the *Kohnstamm phenomenon* (Mathis et al., 1996; Ivanenko et al., 2006; Duclos et al., 2007) known to most inquisitive elementary school students. Ask your friend to stand quietly with the arms hanging by the sides. Then put your hands against the dorsal surfaces of the hands of your friend and ask him or her to press against your hands as strongly as possible trying to lift the arms sideways for 30 s. It is easy to resist the arm motion, so your friend is performing a nearly isometric contraction. After 30 s give your friend a command "relax" and remove your hands: The arms of your friend will slowly fly upward apparently without any specific voluntary command.

The origin of this phenomenon remains a puzzle. Mentioned studies invoked effects of proprioceptive input with or without contribution from efferent processes. A series of studies by De Havas and colleagues (De Havas et al., 2016, 2017, 2018) provided information in favor of a hybrid model interpreting the Kohnstamm phenomenon as a consequence of interactions between efferent and afferent processes. These authors have also come up with a controversial conclusion that this phenomenon is not associated with an efferent copy: Within their model, a hypothetical central generator of the Kohnstamm phenomenon sends its output via the same pathways as those used during the generation of voluntary movements, yet the resulting movement feels involuntary because of the lack of an efferent copy to cancel against the changed sensory inflow. Within the classical definition of efferent copy as a copy of signals from alpha motoneurons, this statement implies that, somehow, during the Kohnstamm phenomenon, collaterals of the axons that carry the efferent copy are selectively inhibited at their targets (e.g., via presynaptic inhibition). As a result, the reafference associated with the movement is presumed to exert the same effects as exafference (see Section 6.2): In particular, it is expected to lead to posture-stabilizing effects moving the arms toward their initial vertical position. This does not happen, however, suggesting that the explanation based on lacking efferent copy is likely incomplete.

Another phenomenon likely reflecting the complex interdependence of efferent and afferent processes is the so-called podokinetic after-rotation effect (Gordon et al., 1995; Weber et al., 1998; Earhart et al., 2002). This effect can be seen after 10–20 min of walking on a rotating platform; when the platform stops and the person is asked to walk in a straight line, the walking trajectory is strongly curved. A similar effect is observed when a person voluntarily turns around while stepping in place (Sozzi and Schieppati, 2016). Under both conditions, the subject does not perceive the effect consciously (Gordon et al., 1995; Weber et al., 1998). The podokinetic effect is not limited to some kind of a spinal-level reorganization of muscle involvement in locomotion as it is accompanied by a consistent shift in perception of straight-ahead direction seen in pointing tasks (Scott et al., 2011). Note that similar effects on the curvature of walking trajectory can be seen under vibration of neck muscles (Bove et al., 2002), a strong stimulus for sensory endings in muscle spindles. These observations support the general idea that unintentionally curved walking emerges at the level of interaction between afferent and efferent processes.

6.6 The concept of iso-perceptual manifold

Let us get back to the earlier example of pressing with a hand against a stop with varying force. This action is associated with changes in the activity of all efferent systems, including the alpha- and gamma-motoneurons, and all peripheral sensory endings. Nevertheless, the arm is perceived correctly as motionless. Along similar lines, co-contracting muscles acting about a joint without moving the joint are associated with veridical perception of the joint being motionless despite changes in the activity levels of all the mentioned afferent and efferent systems. To generalize these mental experiments, everyday experience suggests that a percept may be stable despite possible changes in both afferent and efferent processes as long as they are changed in a coordinated way. This means that, at some level of the central nervous system, there is an *invariant* related to perception of specific body configuration and interactions with the environment. This conclusion can be formalized as a manifold in the high-dimensional space of afferent and efferent variables corresponding to stable percepts of joint positions (and other salient variables). This manifold is addressed as *iso-perceptual manifold* (IPM).

The concept of IPM can be considered at different levels, from a single muscle to the whole body. At any given level of analysis (e.g., those illustrated in Fig. 6.7), relevant afferent and efferent signals form a N-dimensional space, S_N. This space may be viewed as the product of two factors, efferent and afferent, which we will associate here, for simplicity, with two submanifolds with the dimensionalities N_E and N_A. This is a major simplification skipping a few important issues. In particular, variables in the efferent and afferent spaces may not be independent, for example, because of the mentioned phenomenon of alpha-gamma coactivation: A change in efferent signals from alpha-motoneurons is accompanied by a change in activity of gamma-motoneurons, which change the sensitivity of muscle spindle endings to muscle length and velocity and, as a result, affect afferent signals. There is also a possibility that different sensory modalities interact to produce the best estimation of the effector state, e.g., based on Bayesian estimation (Weiss et al., 2002; Körding and Wolpert, 2006).

Earlier, we associated perception with an act of measurement. To measure a set of j salient kinesthetic variables one has to have an origin of coordinates that allows the definition of a j-dimensional vector in the combined afferent–efferent space. Assuming that the percept is stable, afferent and efferent signals can vary within a M-dimensional subspace, which is the IPM at the selected level of analysis. So, during stable perception of an unchanging joint position during muscle cocontraction or during pressing against a stop, all the signals, afferent and efferent, change within the respective IPM. For different percepts, a family of IPMs is presumed to exist within the afferent-efferent space. This family may be constrained by an *a priori* knowledge of the body. For example, some IPMs may not exist because they violate constraints of body integrity (see later).

The IPM structure may be considered as an alternative definition of perception not limited to reported perception (which is so far the only type discussed):

Perception of a physical variable is a reported (including self-reported) consequence of the existence of an IPM in a high-dimensional afferent-efferent space.

The existence of an IPM may affect phenomena of action-perception coupling discussed within the *ecological psychology* approach introduced by Gibson (1979). According to this approach, sensory (afferent) and motor (efferent) processes may be coupled without updating the intrinsic picture of the world. Some of such couplings are trivial, e.g., spinal reflexes, which lead to changes in muscle activation (action) induced by changes in afferent signals (perception) at short time delays. Other phenomena are more sophisticated and involve unconscious adjustments of action while performing another action by the contralateral effector (Sternad et al., 1995) or while observing another person performing a similar action (Schmidt et al., 1990), adjustments of posture in response to motion of the visual field (Dijkstra et al., 1994) or finger contact point in space (Jeka et al., 1998), preparation to impact with an object or surface (Lee et al., 1992), etc. (reviewed in Kugler and Turvey 1987; Turvey, 2007). The concept of IPM may also be relevant to unconscious inference, i.e., an involuntary, prerational mechanism that may be responsible for a variety of phenomena including optical illusions (Wilson and Brekke, 1994; Clifford et al., 2015).

6.6.1 Counting dimensions within the IPM

Counting dimensionalities of spaces and subspaces is a rather complicated enterprise. Let us first consider, within the suggested scheme, the control and perception of a single muscle. If we limit analysis to quasi-static cases, muscle length and force sense ($j = 2$) emerge at the level of interaction between $\{\lambda\}$ to that muscle and all the sources of relevant proprioceptive afferent signals $\{AF\}$. So, the efferent space for a single muscle is one-dimensional ($N_E = 1$). However, the notion of muscle is ambiguous. For example, some muscles are viewed as heads of a larger muscle.

A typical example is addressing soleus, gastrocnemius medialis, and gastrocnemius lateralis as the three heads of a single muscle, triceps surae (cf. the four heads of quadriceps femoris, three heads of triceps brachii of the arm, two head of biceps brachii, etc.). Other muscles are considered as consisting of a few compartments with some degree of autonomy; for example, the two extrinsic finger flexor muscles, flexor digitorum superficialis and flexor digitorum profundus, are considered as consisting of four compartments, each with four distal tendons attached at the individual fingers. Each compartment consists of numerous motor units, and one can introduce λs for individual motor units so that the muscle-relevant λ becomes embedded in a higher-dimensional space of mini-λs for compartments, and in a very high-dimensional space of individual micro-λs specific to individual motor units and alpha-motoneurons.

It is even harder to estimate dimensionality of the afferent space. If each of the main types of afferent signals (e.g., primary and secondary endings of muscle spindles, Golgi tendon organs, articular receptors, etc.) is viewed as a single dimension, then $N_A \geq 3$. Note that this value refers to proprioceptive modalities, not to the number of receptors. In addition, the system of dynamic and static gamma-motoneurons adds to the dimensions of the afferent system, whereas the phenomenon of alpha-gamma coactivation couples the afferent and efferent spaces. Overall, it seems safe to assume that the IPM for perception of muscle length and force is at least two-dimensional.

The joint/digit/limb position sense emerges as a result of interaction between the efferent signals $\{r; c\}$ and afferent signals $\{\alpha; f\}$. At this level, the dimensionality of percepts may be relatively high, corresponding, for example, to perception of position and orientation of the effector as well as to perception of the force and moment vector produced by the effector on the environment. For a single joint with one kinematic degree-of-freedom, $j \geq 3$ corresponding to perceiving joint angle and moment of force as well as the sense of joint capsule tension caused by muscle cocontraction. For a limb, the IPM is potentially higher-dimensional, although estimating its dimensionality is very hard.

At the highest level defined by the task, perception is based on estimating afferent signals $\{S\}$ within the referent frame provided by $\{\boldsymbol{RC}\}$. For example, one may try to estimate the weight and location/orientation of an object lifted by the two arms, or one's body orientation with respect to the vertical, etc. The dimensionality of relevant efferent and afferent spaces, N_A and N_E, would be different in each of those examples. However, it seems safe to assume that the inequality $j < (N_A + N_E)$ holds for typical natural situations.

Note that at each level illustrated in Fig. 6.7, the efferent process defines a reference point, while sensory signals inform on deviations from that reference point. As a result, perception of relevant state variables (a vector $\boldsymbol{P}$) can be expressed as $\boldsymbol{P} = \boldsymbol{RC} + \boldsymbol{S}$. In this equation, the "+" sign should be viewed as symbolic rather than reflecting addition of numbers or vectors. This equation is similar to, and may be viewed as a generalization of, the one proposed by the group of Feldman (Feldman, 2009; Feldman et al., 2013) in the form $Q = R + P$ for perception of muscle length and force.

6.7 Motor equivalence and perceptual equivalence

The term *motor equivalence* has been used for a long time, typically to reflect the ability of animals to perform any given task with variable means (Kelso et al., 1984; Berkinblit et al., 1986; Cole and Abbs, 1987; Wing, 2000; Levin et al., 2003). According to this definition, motor equivalence is a direct consequence of motor abundance at the task level. Everyday examples of motor equivalence include the ability to pick up an object with either hand, to walk to a target along different routes, and to eat a steak cut into small or large pieces. In motor control studies, motor equivalence has been demonstrated using perturbations applied to an effector (or a subset of effectors) from a larger group that performs a task. For example, during pinch grip with the index finger and the thumb, an unexpected mechanical perturbation applied to one of the digits led to a quick modification of the trajectory of the other digit so that the task was performed successfully (Cole and Abbs, 1987). During speech, a perturbation applied to one of the articulators led to adjustments in other articulators such that the sound remained nearly unaffected (Kelso et al., 1984).

Within the framework of the uncontrolled manifold (UCM) hypothesis, motion within a high-dimensional space of elemental variables can be viewed as superposition of two components addressed as *motor equivalent* (ME) and *non–motor equivalent* (nME) (Scholz et al., 2007; Mattos et al., 2011). By definition, ME is motion along the UCM computed for a salient performance variable, which does not lead to changes in that variable. For example, if you touch the tip of your nose with the right index finger, you can move the arm without losing the finger-nose contact. This arm motion is confined to the UCM in the space of joint rotations corresponding to the desired fingertip location in space. Hence, this is ME motion. nME is motion orthogonal to the UCM, and it leads to changes in the salient performance variable. Using the same finger-to-nose example, if you move the finger away from the nose, at least part of this motion is nME.

Fig. 6.8 illustrates the ME and nME motions for the task of producing a certain value of the sum of two elemental variables, e.g., when pressing with two fingers on two force sensors and trying to produce a certain value of total force. If a subject in this experiment starts with a certain total force magnitude (point 1 in Fig. 6.8), then decreases the total force quickly, and then tries to return to the initial force magnitude, he or she will likely end up with another point on the force–force plane. Note that changes in total force correspond to changes from one UCM to another (UCM_1 and UCM_2 in Fig. 6.8). It is not surprising that the subject in this mental experiment ends up close to the original UCM (point 2). It is less trivial that the small nME motion (a small error is unavoidable) is commonly accompanied by a relatively large ME motion.

By definition, if one wants to change a performance variable and this is the only important task component, ME motion is wasteful: It is associated with motion in the space of elemental variables, which requires muscle activations, and does not contribute to the task. Nevertheless, large-magnitude ME motion is observed across tasks and spaces of elemental variables. In particular, imagine that a subject performs

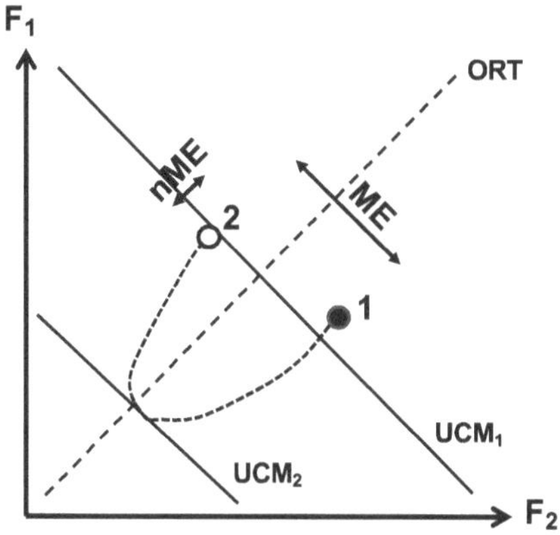

Figure 6.8 An illustration of motor equivalent (ME, along the UCM) and non−motor equivalent (nME, along ORT) motion in the two-dimensional space of two forces produced by two fingers, F_1 and F_2, involved in the task of generating a constant total force. Motion from point 1 to a lower level of total force and back close to the original level leads to point 2; ME > nME.

a pointing movement to a target and an unexpected perturbation is introduced in one of the trials (e.g., one of the major joints is blocked by a rubber band as in Scholz et al., 2000; Mattos et al., 2011, 2013). The subject is asked to point accurately in all trials; so, a quick correction is seen soon after the movement initiation, when the subject perceives an unexpected error in the ongoing action. Corrective movement in the joint angle space may be quantified separately within the UCM for the endpoint coordinate (ME) and orthogonal to that UCM (nME). Such analysis showed much larger ME components (properly quantified given the different dimensionality of the UCM and ORT spaces!) compared with nME components. So, during corrective action, up to 80% of the joint rotations do not correct anything! This is also reflected in much larger ME deviations in the space of muscle activations compared to the nME deviations (Mattos et al., 2013). This result is not a consequence of the anatomy of the human arm: Indeed, similarly high ME deviations have been shown for multifinger force production tasks in isometric conditions when no motion happened (Mattos et al., 2015).

These observations make little sense from the point of view of optimal control. Indeed, as already mentioned, large ME deviations require energy while they do not contribute to the task performance. What could be the origin of those apparently wasteful movement components?

Large ME deviations are natural consequences of the physics of the body, given its abundant design and different stability in different directions within the space of elemental variables. Indeed, consider the scheme in Fig. 6.9. It has three levels. The top level defines how to change a salient performance variable by generating a time profile of the corresponding RC. This level also specifies parameters for the intermediate level, which define stability of the salient performance variable, shown as a matrix of back-coupling gains, $|\mathbf{G}|$ (cf. Latash et al., 2005). The intermediate level maps the task-level RC on RCs for the elements and defines stability of performance

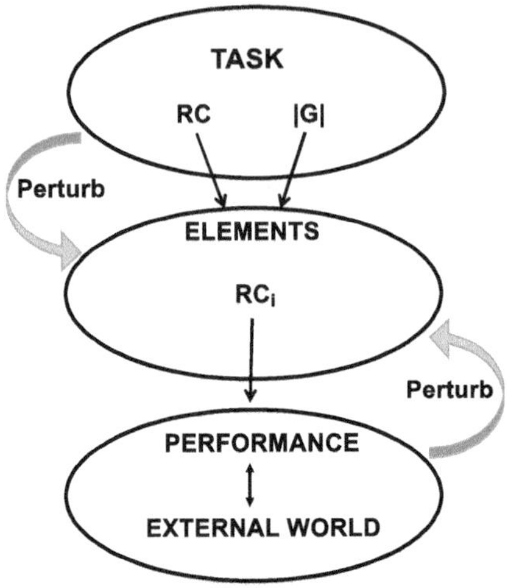

Figure 6.9 At the task level, a value of RC for the salient variable is defined and also stability of that performance variable, shown as a matrix of back-coupling gains, |G|. This input maps on an abundant space of elemental variables where RCs for those variables are defined. This level defines performance in interaction of effectors with external forces. Quick changes in the descending and/or peripheral input into the intermediate level play the role of perturbations and lead to large ME motion.

with respect to the salient performance variable. In particular, at this level, responses of the system to perturbations are channeled primarily into the corresponding UCM. This makes performance stable and defines the typical signature of intertrial variance, $V_{UCM} > V_{ORT}$, described earlier (see Section 4.2). Note that the intermediate level is expected to react similarly to quick changes (perturbations) that come from the periphery and those that come from the top level. In other words, changes in external forces and in descending signals play similar roles of external perturbations for the hypothetical mechanisms that ensure stability of performance. This happens, for example, if the actor decides to introduce a quick correction to an ongoing action without changing stability properties of salient performance variables.

To illustrate this point, consider the following example. Imagine that you took the spring from a typical pen and placed it between the thumb and index finger (you can actually run this experiment on yourself while reading this text). Now try to compress the spring along its main axis very quickly. Let me make a prediction: The spring will buckle and jump away from the hand. Why does this happen? You honestly tried to induce spring motion along its main axis only, but it disobeyed. This is a consequence of the relatively large stiffness and correspondingly high stability of the spring along its main axis and low stiffness and stability in directions orthogonal to the main axis. So, physical systems move not in directions you want them to move but in the least stable directions. This is partly because of the imperfection of any natural actions, which are never directed perfectly along the line of desired motion but have small deviations from the desired direction.

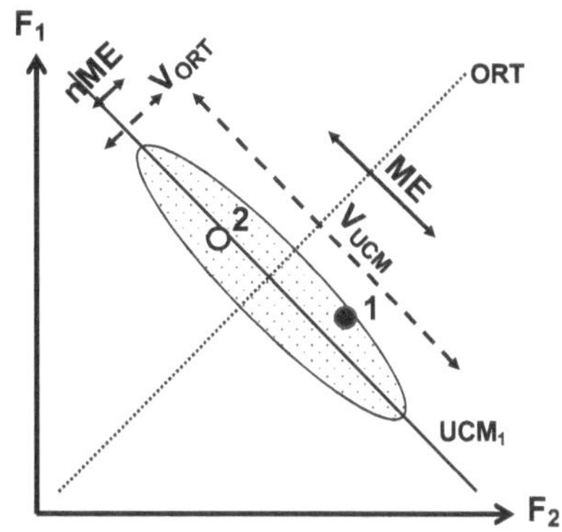

Figure 6.10 The task of fingers producing forces F_1 and F_2, involved in the task of generating a constant total force magnitude. A typical data distribution across multiple trials is shown with an *ellipse* elongated along the uncontrolled manifold (the *solid slanted line*, UCM). Indices of motor equivalence, ME and nME, are expected to correlate with indices of inter-trial variance along the UCM and orthogonal to the UCM space, ORT (V_{UCM} and V_{ORT}) if all points belong to the same normal distribution.

6.7.1 Motor equivalence and structure of variance

The two pairs of indices, ME and nME versus V_{UCM} and V_{ORT}, may be expected to correlate assuming that individual trials can be described as samples from a single normal distribution. This is intuitively clear from the illustration in Fig. 6.10, which is organized similarly to Fig. 6.8. This illustration implies a task of accurate production of a certain magnitude of total force with two effectors; the effectors produce forces F_1 and F_2. If a person performs this task multiple times, the data likely form a cloud elongated along the UCM leading to the inequality $V_{UCM} > V_{ORT}$. Two points are shown within this distribution. The distance between the points along the UCM (which is ME) is expected to be, on average, larger than along the ORT direction (which is nME).

However, the assumption that both points were sampled from the same normal distribution is not trivial! For example, in cases of action correction, measurements made before and after the correction may belong to different distributions. Indeed, a number of experiments, which quantified indices of ME, required the subjects to perform an action with an abundant set of elements, and in some trials the action was unexpectedly perturbed (Mattos et al., 2011, 2015). The subjects were instructed to correct the action as quickly as possible. Those studies documented very large ME motion in the space of elemental variables compared to the required nME motion. After the corrections, however, states of the system could belong not to the same distribution as during unperturbed actions.

In some actions, however, the assumption of a single distribution may be valid, for example, if a person is asked to perform a cyclical change in a performance variable and no perturbations and/or explicit action corrections take place. In such a task, the two variance indices may be computed at comparable phases across the cycles, while ME and nME may be computed between successive cycles, also at comparable phases, e.g., between the peaks. If we assume that each measurement at a particular phase of the cycle represents a sample from a normal distribution, the difference between two random samples will be another normal distribution with the mean $\mu_D = 0$ and standard deviation σ_D. As analysis of motor equivalence consider distances between points, which are always positive, the distribution of the distances (so-called, folded distribution) is characterized by a new mean (μ_X) and new standard deviation (σ_X) (Leone et al., 1961): $\mu_X = \sigma\sqrt{2/\pi}$ and $\sigma_X^2 = \sigma^2 - \mu_X^2$.

The first of these equations suggests that mean ME should be proportional to standard deviation of data points within the UCM, while mean nME should be proportional to standard deviation within the ORT. Such correlations were indeed observed in studies of cyclical whole-body actions (Falaki et al., 2017b). Experiments with quick discrete actions, however, produced a different result (Cuadra et al., 2018). In those experiments, the subjects were asked to produce a quick force pulse with the four fingers of a hand starting from a low initial level of total force and establish the same force level after the pulse. The correlation between ME and square root from V_{UCM} was confirmed, but it was absent between nME and square root from V_{ORT}. The likely explanation is that the subjects corrected their actions within the ORT (as required by the task) but not within the UCM (because it had no effect on performance). As a result, pairs of data points used to compute ME were more likely to belong to the same distribution within the UCM, whereas within the ORT, such pairs were likely to belong to different distributions.

6.7.2 Perceptual equivalence

Consider now the scheme of perception based on the IPM concept. Changes in the high-dimensional space of sensory-motor signals may be separated into two parts. The first one corresponds to no change in perception of an important variable; this means that the trajectory in the multidimensional space is constrained to the corresponding IPM. Such motion can be addressed as *perceptually equivalent* (PE). The second one is associated with a change in perception of that variable (outside the IPM, *nonperceptually equivalent*, nPE).

There are many examples of perceptual equivalence. For example, when a person presses against a stop, there is no illusion of a change in the joint position (Feldman and Latash, 1982a). As mentioned earlier, during this action, all major groups of proprioceptors change their level of activity. Indeed, Golgi tendon organs respond to the increased tendon force, articular receptors react to changes in the joint capsule tension, and spindle endings react to changes in muscle fiber length (which changes despite the fact that the muscle–tendon unit length remains constant) and to changes in activity of gamma-motoneurons. Obviously, motor commands to muscles crossing the joint (λ) change as well. A somewhat simplified interpretation of this effect was offered as resulting from counterdirectional shifts in the two components of muscle length (joint position) sense: A shift in the reference coordinate (defined by λ) is perfectly balanced by a change in the activity level of length-sensitive receptors (Feldman, 2009). This interpretation reduced the multidimensional space of neural

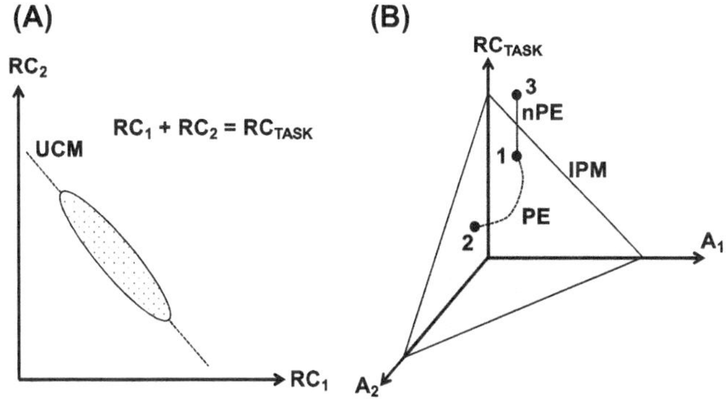

Figure 6.11 An illustration of perceptually equivalent (PE, along the IPM, solid line between points 1 and 2) and non-perceptually equivalent (nPE, outside the IPM, *dashed line* between points 1 and 3) motion in an afferent-efferent space. This is a cartoon system with two efferent elemental variables, RC_1 and RC_2, and two afferent elemental variables, A_1 and A_2. Panel A: typical coordination between the two elemental variables (RC_1 and RC_2) to stabilize the task variable, RC_{TASK}. Panel B: the IPM (the plane) corresponding to perception of a magnitude of the salient performance variable.
Modified by permission from Latash, M.L., 2018a. Stability of kinesthetic perception in efferent-afferent spaces: the concept of iso-perceptual manifold. Neuroscience 372, 97–113. Copyright by Pergamon Press.

processes that contribute to joint position sense to only two one-dimensional processes, afferent and efferent, similar to the earlier equation: $\mathbf{P} = \mathbf{RC} + \mathbf{S}$, without considering the fact that both **RC** and **S** are embedded into high-dimensional spaces. A more complete interpretation would be a shift along the corresponding IPM, i.e., a PE motion illustrated as the dashed trajectory over the IPM from point 1 to point 2 in Fig. 6.11.

Fig. 6.11 is organized assuming a cartoon system with two efferent elemental variables, RC_1 and RC_2, and two afferent elemental variables, A_1 and A_2. For the purpose of illustration on the two-dimensional sheet of paper, this four-dimensional system is split into two panels. Panel A shows the typical coordination between the two elemental variables to stabilize the task variable, RC_{TASK}. Panel B shows the IPM in the three-dimensional space $\{RC_{TASK}, A_1, A_2\}$, which is shown as a plane (for simplicity; it may represent a more complex surface) corresponding to perception of a magnitude of the salient performance variable.

In a classical study, Hulliger et al. (1982) provided evidence for no change in the activity level from length-sensitive spindle endings following voluntary joint movement leading to a change in the length of the corresponding muscle. This was a very much surprising finding given that the subjects in that experiment were well aware of the fact that the joint position changed. How did they know? Within the IPM concept, this result means that the knowledge came primarily from the changed efferent component of joint position sense. This is an example of nPE motion to a new IPM (corresponding to the adequately perceived new joint position) orthogonal to the axis representing activity of length-sensitive endings (illustrated as the solid trajectory from point 1 to point 3 in Fig. 6.11B).

6.7.3 Allowed and disallowed percepts

The abundance of sensory information has important consequences for some of the aforementioned perceptual effects of muscle vibration. Consider a single degree-of-freedom joint crossed by two muscles, a flexor and an extensor (Fig. 6.12). For simplicity, let us also consider that each muscle has one source of length-sensitive afferent information, and these two afferent signals are used for kinesthetic perception of joint position. Of course, this is a two-dimensional projection of a system involving more afferent signals and efferent signals. In the original multidimensional system, for any joint angle, there is an IPM corresponding to veridical perception of the joint state.

Natural joint motion is associated with correlated changes in the length of the two muscles: If one of the muscles is stretched, the other has to shorten. This means that afferent signals from the flexor and extensor muscles, A_F and A_E, have to be linked, for example, by the following equation: $A_F A_E = C$, where C is a constant. For any possible joint position, α_P, there is an IPM that projects on the plane of the two afferent signals illustrated in Fig. 6.12 as a point that belongs to the line defined by $A_F A_E = C$. Across all possible joint positions, all the projections of the corresponding family of IPMs belong to that line reflecting allowed percepts; we will address this line as A_{IPM} (the dashed line in the graph in Fig. 6.12).

Imagine now that high-frequency vibration is applied to one of the muscles, as it is typically done in experimental studies of VIKI, for example, to the flexor; the joint was originally in a state shown by the open circle on the A_{IPM} line in Fig. 6.13. Afferent signal from this muscle, A_F, will change without a corresponding change in the afferent

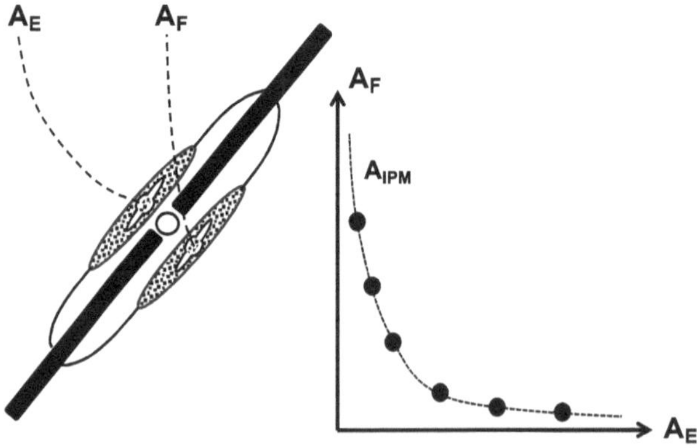

Figure 6.12 Joint anatomical integrity requires that afferent signals from the flexor and extensor muscles, A_F and A_E, are linked: $A_F A_E = C$, where C is a constant. The projections of the IPMs for different joint positions on the plane belongs to the line (the *dashed line*) defined by $A_F A_E = C$. This line is addressed as A_{IPM}.

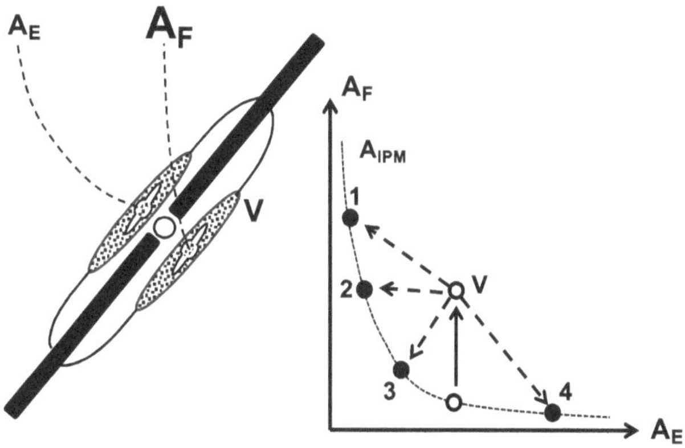

Figure 6.13 Vibration of a flexor muscle leads to higher afferent signals (A_F) from that muscle resulting in a point *(open circle*, V) off the A_{IPM} line (see Fig. 6.12). Projections of this point onto the A_{IPM} line can result in perception of a different joint position corresponding to longer length of the flexor (points 1, 2, and 3), or shorter length of the flexor (point 4).

signal from the extensor (to point V in the graph in Fig. 6.13). Note that this point is off the A_{IPM} line and, as such, it corresponds to an anatomically impossible joint state when the flexor is stretched without a change in the length of the extensor (see also Fig. 6.5). Taken at face value, this combination of afferent signals implies violation of the joint anatomical integrity—a percept that the person's central nervous system refuses to accept (a disallowed percept).

To create an allowed percept of joint position, the point corresponding to a disallowed combination of afferent signals has to be projected onto the A_{IPM} line. Note that this operation leads to a point on the $\{A_F A_E\}$ plane incompatible with the actual combination of afferent (and potentially efferent) signals. This projection can be realized in numerous ways (a few of those are illustrated in Fig. 6.13), and the ultimate percept will depend on where on the A_{IPM} line the new point will emerge. Depending on the location of the projection point on the A_{IPM}, it could be associated with different perceptual illusions corresponding to elongation of the muscle with the vibrator (more likely, points 1, 2, and 3 in Fig. 6.13) or shortening of that muscle (point 4 in Fig. 6.13, as reported in an experiment with inversions of the VIKI direction, Feldman and Latash, 1982c). We do not know what factors define the process of projecting points, corresponding to actual combination of afferent signals, which happen to be off the A_{IPM} line, onto the A_{IPM}. It is possible that this is not a well-defined operation leading to unstable solutions dependent on other factors including those mentioned earlier, such as experience with vibration stimuli, sensory stimuli of other modalities, cognitive task, vibration of the antagonist, etc.

As all afferent signals (as well as all neural signals in the body) are characterized by a certain level of noise, deviations from A_{IPM} curves should happen commonly. Such deviations are probably small and present no major problems for the central nervous system, which is able to execute projections on the closest A_{IPM} leading to sufficiently accurate percepts of salient variables such as joint position, endpoint coordinate in space, etc. Given the phenomenon of signal-dependent noise (Harris and Wolpert, 1998), the deviations are expected to become larger for signals of larger magnitude. For example, given that the primary endings of muscle spindles are sensitive to movement velocity, during fast movements, larger variability in joint position perception may be expected (cf. Cordo et al., 1994).

To summarize, the process of creating a percept may be viewed as search for an allowed IPM closest to the actual set of afferent and efferent signals. These signals may be incommensurate, and metrics of closeness are unknown to us. Moreover, based on some of the mentioned observations, such as changes in the VIKI direction with seemingly irrelevant sensory stimuli and cognitive actions, these metrics may change with other afferent and efferent processes. It is possible that, sometimes, there may be more than one IPM that satisfies the criterion of closeness. In such cases, loss of perception stability may be expected, which is not an unusual phenomenon in studies of the effects of muscle vibration.

6.7.4 Linked perception of kinematic and kinetic variables

There are a few direct consequences of the described scheme of kinesthetic perception. For example, the IPM concept suggests that perception of kinematic variables (such as muscle length, joint angle, and endpoint coordinate) and kinetic variables (such as muscle force, joint torque, and force produced by the endpoint on the environment) have to be intimately linked. Indeed, as illustrated in Fig. 6.4, perceived values of both groups of variables are based on a weighted combination of neural signals from the same sources, peripheral sensory endings, and efferent processes. This implies, in particular, that VIKI should not be limited to illusions of position and velocity but involve illusions of force.

Such illusions have been documented recently in an experiment with accurate finger force production with and without vibration of the finger extensor muscles (Reschechtko et al., 2018; see also Cafarelli and Kostka, 1981). The subjects in that study were asked to produce accurate force levels with one of the hands (the task hand) and then use the other hand to match the perceived force. When the task hand extensors (antagonist muscles for the pressing task) were vibrated, the subjects consistently produced larger forces by the matching hand; when the matching hand extensors were vibrated, the effect reversed and the matching hand consistently produced smaller forces.

These effects are shown schematically in Fig. 6.14A, which illustrates the task as being produced by two muscle groups, flexors (agonists) and extensors (antagonists). This illustration assumes a degree of coactivation of extensor muscles during flexor force production. Vibration of extensors is expected to result in perception of these muscles being longer and producing larger force (point V in Fig. 6.14A). As the extensors were task antagonists, perception of higher extensor force should result in perception of smaller resultant pressing force as compared to the no-vibration condition. So, when the task hand was vibrated, the matching hand (without vibration) was expected to show lower force magnitudes. The observed effects were in the opposite direction. Vibrating the hand performing the matching task is expected

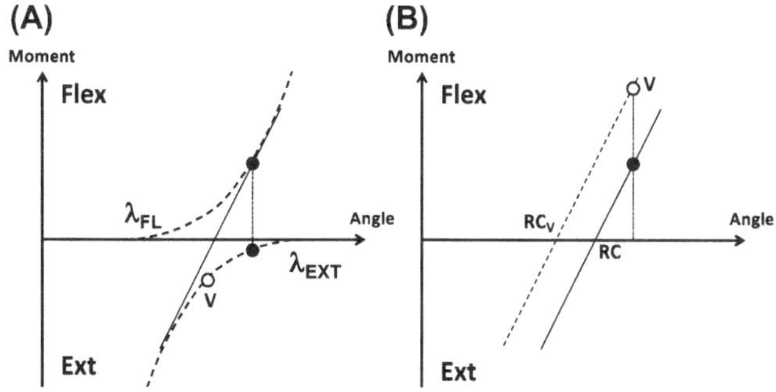

Figure 6.14 (A): Vibration of extensors is expected to result in perception of these muscles being longer and producing larger force (*open circle*, point V, compare to the initial states shown with the two *black dots* connected by the *vertical line*). (B): To account for the perception of larger flexion force (*open circle*, V), RC has to be shifted (to RC_V). Reproduced by permission from Reschechtko, S., Cuadra, C., Latash, M.L., 2018. Force illusions and drifts observed during muscle vibration. Journal of Neurophysiology 119, 326–336. Copyright by American Physiological Society.

to result in its force being underestimated and, therefore, it could be expected to produce larger force magnitudes to match the perceived force produced by the nonvibrated task hand. Again, the experimental results were opposite to this prediction.

Taken together, the mentioned observations suggest that muscle vibration leads to changes not only in the afferent component of kinesthetic perception but also in the efferent (RC) component (as predicted many years ago by Feldman and Latash, 1982a). Moreover, the results suggest that effects of vibration on RC may overpower the expected effects on afferent activity. In particular, under vibration of the extensors, RC value participating in kinesthetic perception shifts toward a more flexed coordinate (RC_V in Fig. 6.14B), which corresponds to overestimation of the actual pressing force (point V), as it was observed in the experiment.

The conclusion on RC shifts under muscle vibration illustrated in Fig. 6.14B has important consequences for understanding action–perception interactions. Indeed, the RC shift induced by vibration should have effects on both perception and action. In the experiment, however, the task hand always had visual feedback and the subjects produced the same force magnitude in the same arm/hand position. How can RC shift, given these constraints? There are two possible explanations. First, it is possible that the subject co-adjusted the *R*- and *C*-commands to the hand (see Section 5.3) in such a way that the vibration-induced change in the *R*-command (from R_0 to R_1 in Fig. 6.15) was accompanied by a co-adjusted change in the *C*-command (from C_0 to C_1) leading to the same force level, shown by point A in Fig. 6.15. But then we have to accept another nontrivial assumption that the *R*-command defines RC for perception, while perception is relatively immune to changes in the *C*-command. This is not a trivial assumption although it fits well the idea that the *R*-command is hierarchically higher than the *C*-command (Feldman, 2015). This latter idea was based on a number of observations including the gradual decrease in the elevated level of

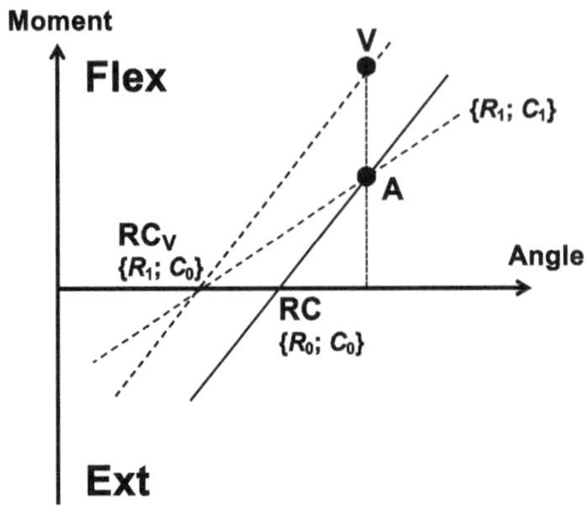

Figure 6.15 In the initial state (point A), the two commands to the effector are $\{R_0; C_0\}$. Shifted RC under vibration (RC_V) corresponds to a new value of the R-command (R_1). To produce the same force, the C-command has to be adjusted (C_1). Note the difference between the pairs of commands, $\{R_1; C_0\}$, involved in the modified perception (point V) and those involved in the unchanged force production $\{R_1; C_1\}$.

agonist–antagonist coactivation after the termination of a fast movement without a change in the final position of the effector.

Another explanation is that central commands for action, such as the *R*- and *C*-commands, differ from their counterparts (their efferent copies) participating in perception. This is not impossible given the dissociation of sensory and motor phenomena in certain neurological disorders and also in healthy persons under certain conditions. For example, motor consequences of muscle vibration are relatively modest. The vibration induces a reflex muscle contraction, the TVR, which takes time to develop and typically leads to relatively low muscle activation levels and forces (De Gail et al., 1966; Eklund and Hagbarth, 1966; Homma et al., 1971). In contrast, perceptual effects of vibration can be very strong leading to perception of very fast joint movements (Sittig et al., 1985) and even impossible joint configurations (Craske, 1977).

6.8 Other well-known and unknown illusions

The concepts of allowed and disallowed percepts illustrated in Fig. 6.12 may be seen as a special case of perception being affected by preconceptions. This topic is definitely beyond my comprehension (BMC!) Nevertheless, let me try to offer an operational definition of *preconception* within the framework discussed in the previous sections. *A preconception is a preexistent A_{IPM}, i.e., a set of allowed (or preferred) IPMs within a high-dimensional set of afferent-efferent signals.* The existence of such an A_{IPM} makes some percepts disallowed, while other percepts may dominate despite the conflicting sensory and/or motor information.

In the field of kinesthetic perception, there are a number of well-known perceptual phenomena that illustrate the important role of preconceptions in interpreting sensory signals. For example, if a person crosses the index and middle fingers of a hand and does not look at the hand, and then another person touches both fingers simultaneously with an object (e.g., a finger), there is a feeling as if the fingers were touched with two objects. The preconception leading to this illusory percept is based on everyday life experience: the radial surface of one of the fingers and the ulnar surface of the other finger cannot be touched with a single small object simultaneously. So, despite the fact that the subject in this imaginary experiment knows that the fingers are crossed, the preconception leads to an illusion.

Another example of the role of preconceptions based on everyday experience is the classical size–weight illusion (e.g., Ernst, 2009; Buckingham, 2014): If a person is presented with two objects of the same weight but different size (e.g., two cubes or two spheres), the smaller object is perceived as being heavier. In this example, the cause for the illusion is violation of the natural expectation based on everyday experience, which suggests that smaller objects are lighter. In these two examples, the sensory signals are not modified by an artificial stimulus (such as muscle vibration) but they are misinterpreted as a result of projection on a preferred subspace in the multidimensional afferent–efferent space where percepts are formed (illustrated as the A_{IPM} in Fig. 6.12).

Several experiments suggest that certain combinations of sensory signals cannot be readily incorporated into the body scheme and lead to motor errors even during movements controlled by the spinal cord. For example, a frog with its spinal cord surgically separated from the brain is able to perform coordinated hindlimb movements to wipe off an irritating stimulus placed on the frog's back (the so-called *wiping reflex* described in the 19th century by Eduard Friedrich Wilhelm Pflüger). The wiping action remained accurate even if the hindlimb was loaded with a heavy bracelet and when motion in one of the major joints of the hindlimb was constrained by an external contraption (reviewed in Latash, 1993; also see Berkinblit et al., 1986). The spinal frog, however, performed consistently erroneous wiping movements, wiping under the stimulus, if a "glove" was placed on the forcefully flexed toes of the hindlimb. Note that limb loading and joint motion constraints can happen in everyday frog's life but toes are never flexed into a fist. Within the current framework, the frog's spinal cord was unable to incorporate the veridical sensory information on the improbable toe position into an image of the body configuration, projected the available sensory information on an allowed A_{IPM}, and performed the wiping as if the toes were naturally extended.

Another illusion following the scheme in Fig. 6.4 has only been reported informally (Latash, 2008; Latash and Zatsiorsky, 2016), although a full-scale experiment was performed in the early 1980s by Anatol Feldman and Mark Latash. The idea of the experiment was to explore position sense of an effector, which moved as a result of direct electrical stimulation of a muscle group. In such a case, no RC change accompanies motion, and the veridical sensory information is expected to be interpreted in a wrong reference frame leading to erroneous perception of final effector position. In the mentioned study, index finger motion was produced by electrical tetanic stimulation of the extrinsic finger flexors, and the subjects used a psychophysical scale to report finger position in a new steady state. The experiment showed that the finger displacement was consistently overestimated corresponding to perception of shorter length of the muscle subjected to the electrical stimulation. These data, as well as the aforementioned data on modifications of the frog wiping reflex (performed by Michael Berkinblit, Anatol Feldman, Olga Fookson, and Mark Latash), have never been published because of the emigration of both Feldman and Latash from the former Soviet Union at the time when the data had already been collected and processed, but before papers could be written.

The concept of IPM implies different stability in different directions within a large-dimensional afferent–efferent space. Indeed, percepts are expected to be highly stable when signals vary within the corresponding IPM and they are expected to lose stability (and, as a consequence, being replaced by other percepts) when signals leave the IPM. In other words, when a person reports identical percepts, the afferent–efferent signals are expected to be within the same IPM. This may be expected, for example, across repetitive trials with exposure to a particular stimulus with the same magnitude or within a single trial when two effectors (e.g., two hands) are subjected to stimuli of the same modality and same magnitude. When the effector involved in such a task is itself composed of several elements, a particular structure of intertrial variance may be expected at the level of elements, somewhat similar to the phenomenon of structured intertrial variance as explored within the framework of the UCM hypothesis (see Section 4.2, reviewed in Latash et al., 2007; Latash, 2008). This may be expected, for example, when perception of total force produced by the hand is viewed as superposition of percepts of forces produced by individual digits.

Imagine that a person is asked to report total force produced by the four fingers of a hand (the task hand) in a pressing task by matching the force with the other hand or using a psychophysical scale. In other trials, the same person is asked to press in the same way by the task hand and then match (or verbally report) the force produced by only one of the fingers. The concept of IPM suggests that percepts of finger forces can vary without affecting accuracy of perception of total force if this variability is confined to the IPM for total force perception. This is an experimentally testable prediction. Indeed, a recent study with force matching has shown larger errors during individual finger force matching compared to total force matching and also compared to single-finger force matching, when both hands produced force with one finger only (Cuadra et al., 2019).

Note that, when the task is limited to force production by a single finger, the IPM for that variable becomes stabilized by co-varied adjustments of afferent and efferent signals relevant to that finger's action. This stability may be lost in a hand performing a multifinger task. This phenomenon is similar to the trade-off between synergies at different levels of a hierarchy described in Section 4.3.1 (see also Gorniak et al., 2007, 2009). Accurate perception of total hand force implies small deviations orthogonal to the IPM, while within-the-IPM deviations may be relatively large. The within-the-IPM deviations may not be aligned with the IPMs for individual finger force perception leading to relatively large errors in perceiving individual finger forces when the task hand performed the four-finger task.

Along similar lines, a person may be expected to show small errors in perception of fingertip position in space as compared to perception of individual joint angles of the arm during the pointing task. To my knowledge, this experiment has not been performed, although informal personal experience suggests that the results will follow the aforementioned prediction. Indeed, close your eyes and place the fingertip of the right hand into a position in space. Now, try to occupy a symmetrical fingertip position with the other arm (of course, with closed eyes). This is easy to do. And now try to match, for example, the elbow (or wrist) joint position between the arms. This is very hard to do despite the fact that you can focus your attention on both elbow joints. Matching joint positions when both arms move one joint at a time is easy.

To summarize, perception of elemental variables involved in a multi-element task is expected to be less accurate than perception of the same variables in a corresponding single-element task or perception of a task-specific variable to which all the elemental variables contribute.

Experimental exploration of the predicted perceptual phenomena may be nontrivial but not impossible either, as shown by a few mentioned experimental studies. Problems with quantification of percepts have already been emphasized. To remind, using psychophysical scales leads to poor resolution and reproducibility, whereas using reproduction of a percept by action involves potentially confounding factors such as the necessity to produce another action, memory, and/or action by another effector. Another problem is that subjects in such experimental studies may focus not on their percepts, as the experimenter hopes, but instead try to match their "commands" (efferent variables, RCs). Several studies have suggested that humans do indeed sometimes "match commands" rather than "reproduce the same magnitude of a positional or force variable" (Rymer and D'Almeida, 1980; van Doren, 1998). But overall this does not seem to be an insurmountable problem.

Stability of Gist

7

Within this part, I am entering a field that is definitely beyond my comprehension (BMC). Indeed, I do not have a clear and unambiguous definition for "gist" or "thought" that would both reflect the intuitive meaning of these words and link them to laws of nature. Nevertheless, I will try to generalize concepts and ideas developed in the previous parts of the book for mechanisms of action control and perception and speculate about their possible relevance to mostly unknown mechanisms involved in language-based communication, decision-making, and other actions commonly addressed as *cognitive*. There are certain common features among all the mentioned processes and actions. In particular, they all may be seen as motion to a target, sometimes a moving target, defined in a corresponding multidimensional space. All of them are characterized by the availability of numerous solutions for typical tasks and stability of salient variables (e.g., those describing the outcome of a process) across those solutions, i.e., the existence of stable solution manifolds in the corresponding multidimensional spaces.

There is evidence from neurophysiological studies that action, perception, and cognition can involve overlapping neural structures and might also share physical (physiological) processes. One large series of studies explored the neural substrate and mechanisms of *motor imagery*, i.e., tasks involving imagining an action without actually performing it. These studies have shown involvement of some of the same brain structures, including primary motor cortex, in tasks involving motor imagery and during the production of the same action (reviewed in Decety, 1996; Jeannerod and Frak, 1999; Mizuguchi and Kanosue, 2017; Mohan et al., 2018). Another highly influential group of studies explored the so-called *mirror neurons* in the brain (reviewed in Rizzolatti and Craighero, 2004; Fadiga et al., 2005), which show activity during observation of an action performed by another animal (person) and also during the production of the same action. Note that in those studies the action was "the same" at a very high, task-specific level, commonly at the level of kinematic topology, which may be associated with cognitive rather than motor processes. For example, the classical studies documenting mirror neurons in monkeys compared the neural activation patterns when a monkey watched an experimenter performing a task and when the monkey performed the task itself. Clearly, the major differences in the anatomy and physiology between the monkey and the experimenter required very different joint motion and muscle activation patterns to accomplish "the same task."

Problems from the field of cognition have been mentioned earlier to illustrate some of the basic concepts typical of the field of motor control (e.g., Latash, 1996). In particular, the task of going to a store and buying a set of products was discussed as an example of the problem of motor redundancy, whereas the redundancy in this case involves perceptual and cognitive components. In this example, the actor selects

Physics of Biological Action and Perception. https://doi.org/10.1016/B978-0-12-819284-9.00007-3

a trajectory from the home to the store based on such factors as the number of street crossings, overall distance, etc., chooses specific products (for example, five apples from a bin of five hundred fruits), selects the fastest cashier line, etc. One can use a formal description of such problems, for example, based on ideas of optimization (see Section 3.1.1). However, intuitively, it seems clear that humans do not select a single optimal solution when performing such tasks. For example, when selecting a desired set of fruits, we are not trying to get to the bottom of the bin, estimate each and every apple according to our criteria, and then pick an optimal set of five apples. More frequently, we allow chance to make selection for us as long as obviously bad apples are avoided. The same is true for selecting a route to the store. Sometimes, when we navigate familiar environment, our mind gets distracted, and we suddenly discover ourselves walking along an unusual route. We do not backtrack to the "optimal route," as long as the current route can be easily adjusted, but rather make necessary adjustments that lead us to the target.

Solutions to such problems may show dependence on history (e.g., selecting the same, routine path vs. being more adventurous and opting for a new route) and on a number of other factors. For example, if the weather in nice, we may select a more picturesque route, even if it is longer and associated with more street crossings. In bad weather, the shortest route is more likely to be selected. Overall, all such problems are associated with "repetition without repetition" (a famous expression coined by Nikolai Bernstein), which means using variable means to achieve the same outcome. To be more exact, in the last phrase, "the same outcome" should be replaced with "similarly acceptable outcomes."

Let me remind that the approach advocated in this book does not have room for computational operations performed within the objects of study, even very complex objects such as the human brain. Therefore, I have been reluctant to accept ideas from the field of optimization that require integration of variables over the time of a future action. Rather, problems are expected to be solved by selecting magnitudes of parameters that define interactions between the object of study and the environment expressed in salient variables. Further in this part, we will try to speculate how this approach, which we call *physical*, can be used for a range of tasks typically viewed as cognitive.

7.1 Encoding–decoding within the body

Using a set of different terms, many phenomena within the animal (including human) body can be seen as a superposition of physiological processes involved in an encoding–decoding sequence. During the encoding step, a potentially important message is distributed over a large number of elements, which can be noisy (e.g., neurons). During the decoding step, information from those noisy elements is combined to generate a salient, low-dimensional variable, which is expected to match the encoded message with acceptable accuracy. The word "message" depending on the context, can mean a thought, a salient percept, a functional action, etc. Although physiological mechanisms involved in all these processes are mostly unknown, their existence seems to be very much feasible.

For example, consider the process of generation of a purposeful action (Fig. 7.1A). Any action starts with a task formulated in terms of values for a relatively small set of task-specific salient variables. These values are used to specify a comparably small set of spatial referent coordinates, $\mathbf{RC}_{\mathrm{TASK}}$ (Section 5.3), which we view as parameters of corresponding biology-specific physical laws. Further, a sequence of few-to-many transformations leads to RCs at hierarchically lower levels, for example, those related to involvement of individual limbs, digits, joints, and muscles (referred to as "elements" in Fig. 7.1). Ultimately, the interactions of RCs for the muscles (which are equivalent to threshold of the stretch reflex, λ) with external forces lead to muscle activation and the production of mechanical variables, i.e., action. This process involves the addition of noise at all steps, for example, due to the unpredictable excitability and threshold properties of individual neurons and may be viewed as encoding $\mathbf{RC}_{\mathrm{TASK}}$ in the noisy signals at the lower levels of the hierarchy. Even more noise is added by the not-perfectly-predictable external forces, which are not shown explicitly in Fig. 7.1. This process seems to corrupt the message carried by $\mathbf{RC}_{\mathrm{TASK}}$ beyond hope.

Further, however, the signals to elements lead to a motor effect (affected by the external force field!), which represents patterns of salient performance variables that typically match closely what was dictated by the task. In other words, the noisy

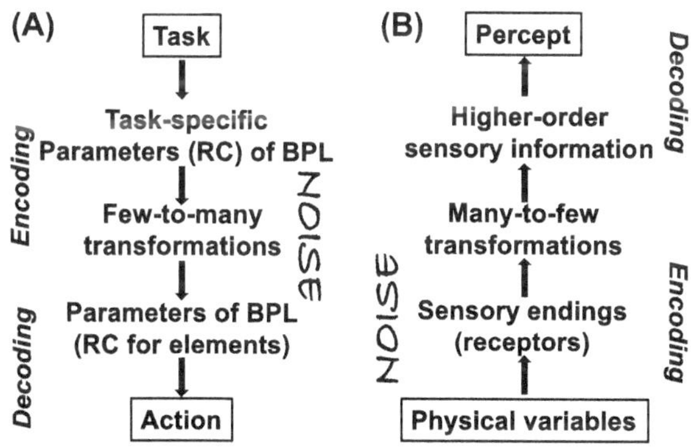

Figure 7.1 An illustration of encoding–decoding sequences in the organization of action (A) and perception (B). (A): Any action starts with specifying a relatively small set of spatial referent coordinates (RC) for task-specific salient variables. A sequence of few-to-many transformations leads to RCs at hierarchically lower levels (limbs, digits, joints, and muscles). This encoding process involves the addition of noise at all steps. RCs for elements lead to a motor effect (affected by the external force field, not shown), which matches closely the task. This decoding is performed by laws of nature that define interactions among body parts and between the body and the environment. (B): Perception starts with a message, values of a relatively small set of salient physical variables, which is encoded in the noisy signals from individual sensory endings. More noise is added by neural signals. The signals are decoded by the central nervous system and lead to veridical perception of the salient variables.

message, e.g., at the level of λs, is decoded and, in the best-case scenario, the action corresponds closely to what has been encoded by $\mathbf{RC}_{TASK}$. This process of decoding is performed not by some smart structures within the central nervous system (no computations are allowed!) but by laws of nature that define interactions among body parts and between the body and the environment.

Along similar lines, perception of salient variables also may be viewed as the process of encoding a message, which represents the actual values of a relatively small set of salient physical variables. This message is encoded in the noisy signals from individual sensory endings (Fig. 7.1B), also affected by the ongoing efferent process (RC values at all the relevant levels). This results in potentially noisy signals at the level of elements at all hierarchical levels of the organization of perception (see Section 6.5 and Fig. 6.7).

Note that encoding in this case is implemented by mechanisms mostly outside the central nervous system, representing interactions between salient physical variables and individual sensory endings. One can view this as projecting a salient external variable on numerous axes in the multidimensional neural space of sensory endings. The projections are nonindependent, and some of them also depend on the ongoing efferent process (e.g., mediated by the system of gamma-motoneurons). However, as a population, the sensory endings provide a powerful mechanism of encoding many salient physical variables at each instant of time, as we know well from our everyday experience, which commonly involves perceiving locations of and interactions with multiple objects.

Further, the signals from sensory endings are decoded with a central role played by the central nervous system and lead to a veridical, or close to veridical, perception of the salient variables. For example, a change in the location of the fingertip with respect to the body following a voluntary arm motion leads to changes in numerous afferent signals from proprioceptors and is also accompanied by changes in efferent processes involving numerous neural cells, including gamma-motoneurons. All the mentioned individual signals are noisy. Nevertheless, the fingertip position can be perceived reasonably accurately, which may be viewed as a result of successful decoding of information on the fingertip location that has been encoded using the numerous involved noisy elements.

Neural signals related to functionally important mechanical variables that do not have dedicated peripheral sensory endings can be found at subcortical levels. For example, recording from the spinocerebellar tracts in cats during locomotion showed signals not reflecting individual muscle length, muscle force, and joint position time profiles, which seem to have dedicated peripheral sensory endings, but global variables such as leg length and orientation (Bosco and Poppele, 2002). Note that no proprioceptors generate signals reflecting these two variables, and they have to be extracted from activation levels of sensory endings sensitive to local, elemental, variables such as muscle length and velocity, tendon forces, and rotation in individual joints. So, the decoding process is likely distributed among multiple structures in the body and, at least partly, proceeds at the spinal cord level.

In a perfect world, the result of encoding–decoding leads to a close to veridical reconstruction of the original message, which is resistant to noise and partial loss of information at intermediate steps. Experiments with joint fixation and hindlimb loading during the wiping reflex in spinal frogs (Berkinblit et al., 1986; Latash, 1993) have shown that this is true for motor processes guided by sensory stimuli even when such problems are solved by the body controlled by the spinal cord only. On the other hand, observations of unchanged joint position sense in patients with total joint replacement (Cross and McCloskey, 1973; Karanjia and Fergusson, 1983) confirm this conclusion for perception. Sometimes, however, there are consistent motor errors (for example, during rotation of a person in a centrifuge, Lackner and Dizio, 1994; DiZio and Lackner, 1995) and kinesthetic illusions (for example, during muscle vibration, Goodwin et al., 1972; Lackner and Levine 1979; Roll and Vedel, 1982), which show limitations of the decoding process. This process seems to be resistant to some, limited, distortion of information at the level of elements but is unable to handle very unusual situations, which are not common during natural activity of the animal.

7.2 Gist–Text–Gist transformations

In this section, I assume that at least part of a thought can be adequately expressed verbally. This is, by itself, a strong assumption. A great Russian poet of the 19th century, Fyodor Tyutchev (1803–1873), wrote in one of his most famous poems "Silentium!" the following line: "A thought once uttered is untrue" (translation by Vladimir Nabokov). Indeed, quite commonly, we are at a loss for words and, after speaking up, find ourselves dissatisfied with the produced text. Let us leave aside the deeply philosophical idea that any thought is untrue, at least not completely true, even without being uttered. The issue is not whether a thought is true, but whether a text can be an adequate expression of this, potentially false, thought. Based on the aforementioned quotation from Tyutchev and introspection, I am going to assume that only part of a thought, addressed further as its *Gist*, can be adequately expressed using natural human language. More subtle aspect of any nontrivial thought, which is likely to be lost in the process of converting them into text, remains beyond the current discussion (and BMC!).

Hence, assume that all human communications involving language are based on Gist–Text–Gist transformations. One person (the Speaker; could be the Writer, but let us not complicate the story) produces a phrase (Text), commonly accompanied by nonverbal cues, with the purpose of inducing an effect in another person (the Listener). Assume that the Speaker's intention can be associated with a certain value, likely multidimensional, of Gist of his/her message, Gist-S. Gist-S is encoded in a Text (and nonverbal cues), which is then interpreted by the Listener as Gist-L. Sometimes, the intention of the Speaker is to make sure that Gist-L is identical to Gist-S. Of course, Gist-S may have only indirect relation to the produced Text.

For example, in some countries, when you ask for price at the market, the seller gives you a completely crazy number. Assuming that the seller considers you a reasonable person, not someone ready to pay anything, the purpose of that first statement is to invite you to negotiate (Gist-S = "Let us talk"). In fact, if you agree and pay the asked price, the seller may be upset, even offended! Experienced buyers interpret the initial statement adequately and enter the negotiation process. They will respond with an offer that may be two orders of magnitude smaller than the asking price (see, for example, the market scene in my favorite movie of all times, "Casablanca"). The seller may pretend being offended, but this is part of the performance. A lengthy negotiation will follow and, typically, result in the object being sold for a reasonable price. You may ask: "Why did not the seller ask for the reasonable price from the very beginning? This could save a lot of time for both." True, but then the process of human communication would be all but lost, and this process may be at least as valuable for the participants as buying and selling is. To quote the famous Tevye, the Dairyman, the answer is *Tradition*!

When a sulking child comes to the Mom, the text the child utters may have little direct meaning; Gist-S = "I am unhappy and want to be comforted, fed, and/or put to sleep." These, and many other, examples illustrate that the direct meaning of a text may have little to do with its Gist, which depends on the context, age, cultural background, knowledge of the language, and numerous other factors.

In texts produced by professional poets and writers, quite frequently, the direct meaning plays a secondary role. In one of the small funny stories, "The Poet," by a classic of the Czech literature, Karel Čapek (1890-1938), a poet witnesses a car accident: In a village, a car hits an elderly lady and then drives away. The poet is the only witness and when asked direct questions about the car license plate, color, etc., the poet is unable to produce any coherent information. But then he writes a poem about a beautiful young girl getting into an accident, and in that poem, he has lines that nobody, including himself, could understand: "swan's neck bosom drum and cymbal, why do I weep so much." The very smart village police officer assigned to the case immediately deciphers this text as suggesting that the car license plate was 235. Indeed, 2 looks like the swan's neck, 3—like a bosom (sideways), and 5—like a drum and the cymbal above it. And, lo and behold, they do find a car with exactly that license plate and prove that it was involved in the accident!

I have my own example of using context as an effective communication tool unrelated to the actually uttered text. In one of the European countries, local people do not like foreigners who speak their language with errors and/or an accent. They also dislike those who speak English. (Note that I have been very much politically correct and avoided naming the country!) Unfortunately, my command of the local language was rather poor. As a result, I had a few unsuccessful attempts at communicating with inhabitants of that beautiful country in typical situations such as asking for directions, buying something, or ordering a meal. They pretended that they could not understand either my broken Local or my acceptable, albeit not native, English. Then, I developed a strategy. I would come to an unsuspecting local person and start speaking in my very fluent native Russian! The text did not matter; its Gist-S could be summarized as "I am from a weird country and do not speak Local." The local person got confused and tried to find some means of communication, gesturing, speaking slowly, etc. And then I switched to English. What a relief! All locals suddenly acquired an ability to understand and speak the previously ignored English reasonably well.

All these examples suggest that Gist-Text-Gist transformations are very complicated (definitely BMC!). For the purposes of this book, let us consider a simplified situation when the Speaker's intention is to encode a specific Gist-S in such a way that the message decoded by the Listener (Gist-L) is identical, or at least very close, to Gist-S. This makes the ideal Gist-Text-Gist transformation a unity transformation. The Gist-Text-Gist transformation may be viewed as two processes: First, encoding a message, using verbal and nonverbal means, which are likely to add a lot of noise to the intended message (including, but not limited to, "mmm," "well," "you see," etc.), and second, decoding it. Note that the first process may be purposefully misleading when the Speaker wants the Listener to believe something that is not true. For example, looking excessively anxious during an interrogation process may be purposefully faked to make the interrogator believe that what the person is saying is false, while it may be 100% correct. The message may have both informational and emotional components, and the latter may be more important than the former, in particular, in pieces written by professional writers and poets or uttered by professional actors. But, as already mentioned, even a child sometimes says something not to deliver direct information apparently encoded by the text but to complain and/or to ask for sympathy.

As mentioned earlier, encoding–decoding is inherent to all the processes within the body. The examples from the fields of action and perception suggest that encoding–decoding represents an effective means of preserving essential features of a message, its Gist, despite possible distortions in the transmission process. During language-based communication, the Speaker can encode a desired message using multiple combinations of verbal and nonverbal means. The Gist of the message is typically stable, which means that it can be decoded by the Listener even if the communication channel distorts some of the sounds, possibly making some of the uttered words incomprehensible. This can happen, for example, during a telephone conversation when the connection is poor. Having nonverbal cues is usually helpful to ensure stability of the Gist, for example, when the Listener can see the Speaker as during face-to-face conversations or video conferencing over the Internet. But even if the physical signals during the communication process are not distorted, the natural variability of phrases and prosody can be seen as a source of noise used at the encoding stage.

Stability of meaning (a term that is not typically defined, but I am going to assume that it is synonymous with Gist) has been explored in a number of studies (e.g., Raczaszek et al., 1999; Liu et al., 2013), which emphasized, in particular, the role of recent history and prosody. Prosody is particularly important and can change the meaning even if the sentence is strongly semantically biased (Misono et al., 1997; Blasko and Hall, 1998). In other words, the meaning of a written phrase can lose stability and be replaced by a different meaning if the phrase is spoken (see the earlier quotation from Tyutchev). It has also been suggested that diversification of languages could happen at the level of meaning of lexical items, which could lose stability leading to frequency-dependent selection (Pawlowitsch et al., 2011).

Each of the two steps, Gist-S–Text and Text–Gist-L, allows numerous solutions that depend on the level of Speaker's and Listener's command of the language, overall level of education, concentration, and a few other factors. Clearly, if one does not know the language, the Text–Gist-L transformation would be impossible or lead to a trivial default result (Gist-L: "The Speaker speaks a language that I do not know"). If a person has not read certain books, both using and deciphering hidden or explicit quotations from those sources become impossible. There are other examples of miscommunication, e.g., in cases of a special text (e.g., a set of mathematical theorems) delivered to a nonspecialist. This happens when the Speaker misjudges the level of preparation of the Listener—a common mistake at seminar presentations during job interviews. In such situations, when the Speaker assumes too much knowledge on part of the audience, important aspects of Gist-S can be lost, resulting in a trivial, but correct, Gist-L: "The Speaker cannot speak at a level understandable by the audience." For effective communication, one has to follow a simple rule: The intended message has to be understandable by the audience, and it has to be encoded in a way that the audience can decode successfully, i.e. with highly stable Gist that can withstand the natural variability among the Listeners.

7.3 Systems of coordinate: the Iso-Gist Manifold

We do not know what coordinate systems are used to create phrases that encode specific messages. Likely, these coordinate systems are very high dimensional and we spend the lifetime learning and refining them. They may have links to basic words, notions, and modifiers, including those encoded in prosody and nonverbal means of communication such as mimics and gestures. For the sake of simplicity, imagine that the problem of creating the Gist-S–Text transformation can be expressed using two large groups of variables. One of them is related to information (Gist) that has to be encoded, whereas the other—to Text used in the process of encoding. Fig. 7.2 illustrates possible relations between those two large groups of variables using only one axis per group. Gist-S in this example corresponds to a point on the Gist axis. There are potentially an infinite number of actions (Texts) adequate to convey any given value on the Gist-S axis. They all correspond to points on a vertical line, which is in fact a representation of a multidimensional space on this two-dimensional diagram. We will address this multidimensional subspace as Iso-Gist Manifold (IGM).

Consider, as an illustration, a purposefully ridiculous piece by a great Russian writer and artist, Vagrich Bakhchanyan, written in 1974 (V.Bakhchanian, "Mukh Ujma [Khudozhestva]", 2006). Here comes a brief excerpt from that piece with a speech supposedly delivered by the General Secretary of the Communist Party, Leonid Ilyich Brezhnev:

> *"Dear comrades!*
> *Soon our country will have a lot of Pepsi (applause);*
> *Our country will have a lot of Pepsi soon (applause);*
> *A lot of Pepsi our country will soon have (applause);*
> *Our Pepsi will a lot of country have soon (applause);*
> *Our soon Pepsi country will have a lot of (applause);*
> *Of Pepsi country have a lot our soon will (applause)…"*

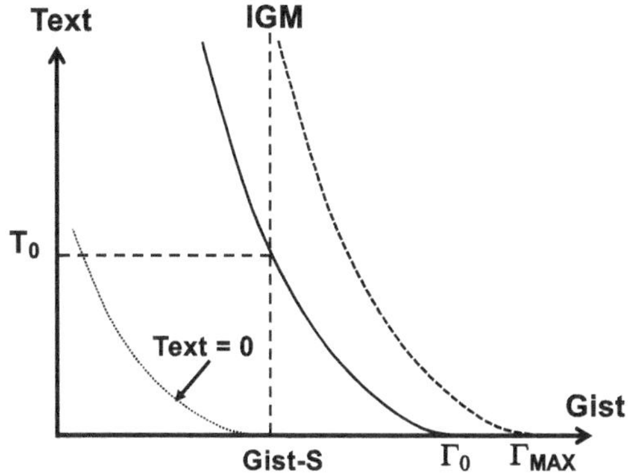

Figure 7.2 Possible relations between Gist and Text. Speaker wants to express Gist-S, a point on the Gist axis; all the points on the *vertical dashed line* correspond to Texts adequate to reflect Gist-S. This multidimensional subspace is the Iso-Gist Manifold (IGM). A value of a parameter Γ defines a relation between Gist-S and Text illustrated with the *solid curved line* for $\Gamma = \Gamma_0$, which produces a text for Gist-S (e.g., T_0). Maximal value of Γ (Γ_{MAX}) is related to the maximal value of Gist the Speaker can convey. Low values of Γ may be unable to express Gist-S (see the *dotted line* with Text = 0).

The original piece in Russian is actually quite a bit longer with more word permutations (the author took the liberty of converting the original text into English). Note that, compared with English, the Russian language is much more forgiving with respect to the word order. Of course, the IGM is not limited to word permutations and contains many more Texts that convey the same message (have the same Gist value). Realization of specific solutions depends on such characteristics of the Speaker as command of the language, level of education, vocabulary, being focused on the task, emotional state, and potentially many others.

Fig. 7.2 illustrates a reflection of this set of Speaker's characteristics with a point on the Gist axis (e.g., Γ_0). This value is a parameter that defines a relation between Gist and Text illustrated with the solid curved line in Fig. 7.2. Its maximal value (Γ_{MAX}) is related to the maximal value of Gist the Speaker can encode successfully, corresponding to a Text(Gist) dependence shown with the dashed curve. For anything more sophisticated (larger Gist values), the Speaker is unable to produce an adequate text. For example, if the Speaker does not know a particular language well, he or she can only create phrases describing relatively simple messages, and, even for those messages, the number of possible adequate phrases may be small. If the Speaker does not know a certain special field (e.g., mathematics), Gist values to the right of Γ_{MAX} are beyond the Speaker's ability to express verbally. Note that the Gist axis is a representation of a multidimensional space, and the expression "to the right of Γ_{MAX}" implies a subspace in that multidimensional original space corresponding to Gist values that are beyond the Speaker's abilities to express.

In a sense, Γ may be viewed as analogous to the concept of λ within the equilibrium-point hypothesis (see Section 5.1 and Feldman, 1966, 2015). It represents threshold for the Speaker's ability to produce an acceptable Gist-S—Text transformation, just like λ represents the length threshold for the muscle to be able to produce active force. Therefore, the concept of Γ brings the scheme illustrated in Fig. 7.2 into the group of threshold-control schemes.

For a given Gist-S, selecting different values of Γ larger than Gist-S produces a range of specific solutions represented in Fig. 7.2 by the intersection of the corresponding Text(Gist) curve and the IGM. Realization of a specific text (e.g., T_0 in Fig. 7.2) may be affected by a host of factors, outside this simplified scheme. In particular, subtle aspects of the intended meaning (those beyond its Gist and BMC) may force the Speaker to use different values of Γ, between Gist-S and Γ_{MAX}, (e.g., Γ_0 in Fig. 7.2), leading to different Text values to express a given Gist-S. For example, if you tell a story several times to different friends, the actually produced texts will vary due to the selection of varying values of Γ, whereas the Gist of the story might remain unchanged. The ordinate in Fig. 7.2 is organized from simpler (minimalist) texts corresponding to lower values of Text, which can be produced by setting Γ just over Gist-S, to more sophisticated texts corresponding to higher values of Text, which require higher values of Γ. For $\Gamma <$ Gist-S, there is no solution, i.e., no adequate text (Text $= 0$; see the dotted curve in Fig. 7.2).

For a given Speaker, Γ_{MAX} defines maximal Gist value the Speaker can convey under perfect circumstances involving maximal concentration. It also defines a range of available solutions for a given submaximal Gist: As it is clear from Fig. 7.2, larger Γ ranges afford larger subspaces within the IGM available for Gist-S—Text transformations. The actually used magnitudes of Γ (up to Γ_{MAX}) may be defined by the current state of the Speaker, which includes such factors as level of alertness, concentration, involvement into the task, etc.

For example, the goal of a seminar or a lecture may be viewed as presenting the audience with a time function Gist(t). This goal may be achieved with various Γ(t) functions (for example, when teaching the same course multiple times), and the actual presentations by the Speaker may vary within a broad range. If the Speaker is alert and well prepared, the audience is responsive, and there are no distracting events, nearly the whole range of Γ values may be used up to Γ_{MAX}. In most situations, however, the range of Γ is limited by values smaller than Γ_{MAX} resulting in not-so-brilliant, but still adequate, presentations. Of course, we assume that the Speaker is trying in each attempt to express a Gist(t) function, not to memorize a text and repeat it as a parrot, or, equivalently, to read a written lecture.

Imagine that a particular solution has been realized resulting in a coordinate T_0, which reflects both verbal and nonverbal means of expression. The next step in the communication process is to decipher T_0, i.e. to convert it into Gist-L.

The process of decoding T_0 by the Listener is illustrated in Fig. 7.3. Here the X-axis corresponds to the incoming message (Text). If the Listener is not sophisticated enough, for example, if he or she does not know the language, Gist-L value is going to be zero (or a default value unrelated to Gist-S). Simply knowing the language may

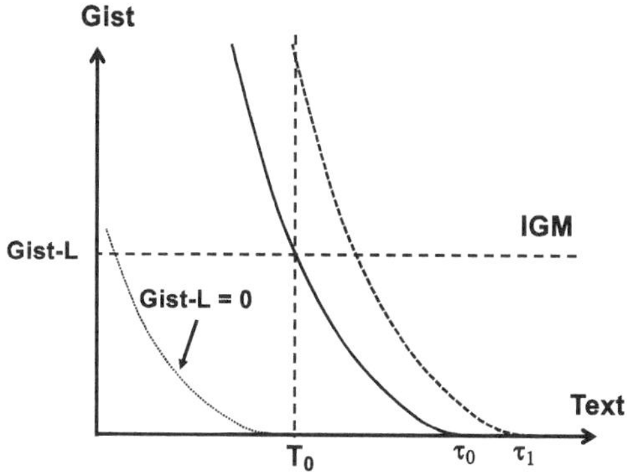

Figure 7.3 The process of decoding T_0 by the Listener. A parameter τ defines a Gist-L(Text) relation (*solid curve*). Being unable to understand the text is illustrated with the *dotted line* ($\tau < T_0$). Gist-L can have a higher value than Gist-S (illustrated with the *dashed curve* with $\tau_1 > \tau_0$).

be sufficient to extract part of the encoded Gist-S but insufficient for extracting its subtler aspects. These may include aspects of Gist-S understandable only to specialists in the field, understanding hidden and explicit quotations, allusions, hints, etc. So, Gist-L extracted from T_0 varies with the Listener's level of education, sophistication, and attention.

This is illustrated in Fig. 7.3 with a curve originating from a threshold value on the X-axis (τ). For values of τ below T_0 (the dotted line), no specific information can be extracted. We ignore the aforementioned trivial case of Gist-L = "the Speaker speaks a language that I do not know" and also cases when the Listener pays no attention to the Text, Gist-L = "I do not care what the Speaker is talking about." An increase in the distance between τ and T_0 allows the Listener extracting more and more information from the message. In fact, sometimes Gist-L can have a higher value than Gist-S (illustrated with the curve with $\tau_1 > \tau_0$), reflecting not only the message encoded by the Speaker but also such features as the Speaker's origin (based on the accent), level of education (based on the choice of words and possible grammatical mistakes), truthfulness of the message (based on advance knowledge, prosody, and nonverbal cues), etc. This was very well illustrated by Mr. Sherlock Holmes and Prof. Higgins in the respective classics.

Consider now the case when Gist-L is identical to Gist-S. T_0 is a point within the appropriate multidimensional IGM defined by the selected value of Γ, Γ_0. Selecting different values of Γ produces different T values for the same Gist-S (a family of $T_{0,\Gamma}$, which can be referred to as T_{IGM}). This family of Texts belongs to the same IGM and, after proper decoding, all these Texts should lead to the same Gist-L. Given the two-dimensional piece of paper, it is rather hard to visualize how different values of $T_{0,\Gamma}$ can be decoded successfully by the Listener as resulting from the same Gist-S. One has to have at least two axes for Text corresponding to Texts that are all adequate to express the same Gist (within the corresponding IGM, T_{IGM}) and those that reflect different Gist values (outside a single IGM).

Within this scheme, Gist encoding and Text decoding processes are controlled using threshold variables, Γ and τ, which are both somewhat similar to λ (Feldman, 1966, 1986; see also Section 5.1). There are several potentially important differences, however. In particular, λ, which reflects neurophysiological processes in the body (such as stretch reflex), is defined primarily by the evolution. In contrast, Γ and τ reflect both the lifetime person's experience and the current state including level of alertness, concentration, etc. In other words, λ is a global parameter of a law of nature, common for all skeletal muscles across individuals and species, whereas Γ and τ are parameters within laws of nature that are individual-specific. This does not disqualify them from being laws of nature. Indeed, many classical laws of nature are applicable to specific groups of objects; for example, the Hooke's Law is applicable only to deformable objects that generate force against deformation and accumulate potential energy (addressed as "springs"). A living system is a Universe of its own, much more complicated than the class of springs, and there seems to be no reason to disallow it to develop its own laws of nature based on both genetic material and lifetime experience.

Generating a message typical of the process of communication can be illustrated with a trajectory in the space of elemental variables (EVs) reflecting verbal and nonverbal means of communication from an initial state to a corresponding IGM. Fig. 7.4 illustrates such possible trajectories to three different solutions within an

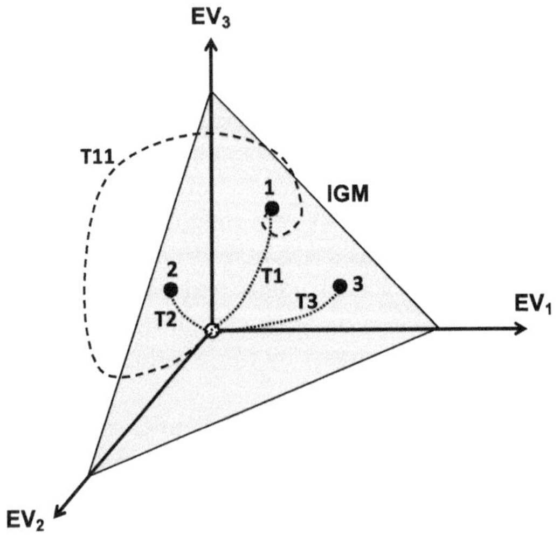

Figure 7.4 Generating a message can be illustrated with a trajectory in the space of elemental variables (EVs) from an initial state to a corresponding Iso-Gist Manifold (IGM). Three such possible trajectories (T1, T2, and T3) to different solutions within an IGM (points 1, 2, and 3) are shown assuming that the subject started from a state with all EV = 0. This figure also illustrates a case when a trajectory goes way outside the desired IGM and needs a correction to lead to an acceptable solution (text T11).

IGM (points 1, 2, and 3) assuming that the subject started from a state with all EV = 0. This figure also illustrates a case when a trajectory goes way outside the desired IGM and needs a correction to lead to an acceptable solution (trajectory T11). As a lecturer, sometimes, in the middle of a lecture, I suddenly hear my own voice and think in panic: "What am I talking about?!" In most such cases, however, being caught on a wrong trajectory allows correcting the path and reaching the desired IPM before anybody in the audience realizes what has just happened.

7.3.1 Moving to an Iso-Gist Manifold: the role of preconceptions

The process of deciphering a message, i.e., turning Text into Gist-L, may lead to unexpected outcomes similar to those described earlier for illusory percepts (Sections 6.3 and 6.8). Preconceptions (assumptions) can change meaning of words (e.g., Collins and Robinson, 2005) based on recent history and other factors leading to changes in the perceived meaning of the Text. This may be seen as projection of an actual Text onto a preconceived IGM corresponding to what the Listener views as feasible. In the discussion of kinesthetic perception, Figs. 6.12 and 6.13 visualize such preconceptions as a line in the space of afferent signals corresponding to anatomically feasible changes in the length of opposing muscles crossing a joint, A_{IPM}. Any deviations of signals generated by peripheral sensory endings from those that belong to an A_{IPM} curve are viewed as impossible and replaced by a point on A_{IPM} leading to a false, but subjectively acceptable, percept. Similar effects can happen in the space where IGM is formed corresponding to a hypothetical acceptable subspace, which we will address as A_{IGM} for consistency.

Effects similar to those illustrated in Figs. 6.12 and 6.13 may be even more important in cognitive processes such as speech generation and comprehension. Knowing about a person and/or situation can make deciphering text easier if it fits expectations based on the preexistent knowledge. In such situations, Gist-S can be extracted based on fragmented and/or highly noisy information. On the other hand, if the text contradicts expectations, extraction of Gist-S may be nontrivial, and it can be replaced by a nonmatching, false Gist-L.

Fig. 7.5 illustrates the space of two EVs within the high-dimensional space where Text is defined, EV_1 and EV_2, with two curves (*a* and *b*) corresponding to allowed variations of those two variables. Note that these variations belong to different IGMs so that points along each of the curves correspond to different Gist values. Imagine that a Listener receives a Text, which involves values of the two variables that do not belong to either A_{IGM} and, hence, are viewed as impossible (T_{IMP}). For example, they represent a major violation of the rules of grammar or logic. To make the message interpretable, some kind of projection onto an A_{IGM} is required (four such projections are shown in Fig. 7.5) with two potential outcomes. First, projecting a T_{IMP} point onto an A_{IGM} can be seen as stabilizing Gist-L corresponding to this A_{IGM}; as a result, minor mistakes such as slips of tongue or minor grammatical errors do not lead to misinterpretation of the encoded Gist-S. Second, a major deviation

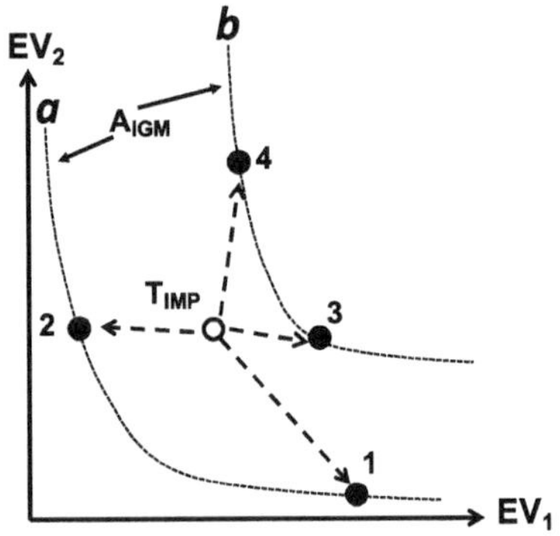

Figure 7.5 Variations of two elemental variables (EV_1 and EV_2) within the high-dimensional space where Text is defined are allowed to covary along specific curves, A_{IGM} (two such *curves* are illustrated, a and b) reflecting different Iso-Gist Manifolds (IGMs). If an incoming Text corresponds to an impossible combination of EV_1 and EV_2 (T_{IMP}), this point is projected on an allowed curve, thus stabilizing the corresponding Gist value. The possibility of projecting on different curves leads potentially to instability of the extracted Gist.

from any A_{IGM} may result in projecting the Text onto a preconceived A_{IGM} unrelated to the original message and potentially lead to misunderstanding. The aforementioned effects of prosody on stability of meaning may represent examples of such a mismatch between the text and the accompanying prosody leading to loss of Gist stability and a switch to another A_{IGM}.

Consider, for example, messages containing grammatical and stylistic errors typical of child speech. Such Texts can be understood adequately because they belong to A_{IGM} given that the Speaker is a child. If similar errors occur in Texts generated by an adult native Speaker, they become T_{IMP}, and their Gist-S may be replaced with a different Gist-L corresponding to a neighboring A_{IGM} or even with a default: Gist-L = "This person makes no sense; he or she is impaired, possibly drunk."

7.4 Stability of sound versus stability of Gist

Stability of Gist is a crucial component of successful communication, whereas stability of sound produced during speech is an important, although not the only one, contributor to stability of Gist.

It has been traditionally accepted that the main and maybe even the only goal of the neural coordination of articulators is to produce a certain sequence of sounds (Johnson et al., 1993; Nieto-Castanon et al., 2005). Several studies used mechanical perturbations applied to one of the articulators and documented very quick (at a delay of about 20 ms) responses in other articulators (Kelso et al., 1982, 1984; Abbs and Gracco, 1984). The observed patterns of those quick responses, possibly mediated by the motor cortex (Ito et al., 2005), have been interpreted as reflections of mechanisms stabilizing certain salient features of the produced sound by adjustments within the abundant group of articulators. Similar conclusions have been drawn based on studies of covariation of the lower and upper lip motion during speech movements in both natural conditions and in somewhat unusual conditions, in particular, with a bite block (Folkins and Linville, 1983).

A more recent series of studies challenged these conclusions (Tremblay et al., 2003; Nasir and Ostry, 2006). In those experiments, a small mechanical perturbation to articulators produced short-latency corrective actions although the perturbations did not lead to any perceivable changes in the acoustical signal as evaluated by experienced listeners. These observations suggest that speech actions may not be limited to achieving a particular acoustical goal but also have another, possibly somatosensory, goal that is reflected in the response of articulators to such small perturbation.

Other studies explored actions by articulators during natural speech in the absence of external perturbations. For example, varying the phonetic context produced local coadjustments in the articulator kinematics, while varying the speaking rate affected globally the whole utterance (Shaiman, 2002). Several studies of speech used measures reflecting variability among EVs and tried to link it to stability of performance (acoustic stability, Nieto-Castanon et al., 2005). A number of studies of the trajectories produced by articulators have suggested that multiple motor equivalent solutions (see Section 6.7) could be used for the production of the same acoustical effect (e.g., Perkell et al., 1993). Likely, these motor equivalent solutions represent motion within the uncontrolled manifold (UCM, see Section 4.2) associated with the desired acoustical effect. Note, however, that the IGM corresponding to a natural utterance may involve nonmotor equivalent solutions with respect to stable acoustics. The same Gist may be reflected in numerous phrases (acoustical realizations), as long as they all belong to the corresponding IGM.

The relations between the concepts of UCM and IGM are somewhat similar to those between the UCM and IPM. On the one hand, within the UCM, motion (motor equivalent motion) does not lead to changes in salient percepts or Gist, whereas intertrial variance within the UCM (V_{UCM}) is a sign of stability of the corresponding action and associated percept or Gist. On the other hand, both IPM and IGM involve dimensions outside the corresponding UCM, and, as a result, nonmotor equivalent motion can be perceptually equivalent or Gist-equivalent, whereas high variance outside the UCM (V_{ORT}) may not mean destabilization of the corresponding percept and Gist.

Perception of speech has been viewed as a special case of multisensory integration (Massaro, 2004). Whereas visual and auditory sources of information typically reinforce each other, sometimes they come into conflict. Maybe, one of the best-known examples is ventriloquism, when the synchronized motion of the lips of the doll, in the absence of motion of the lips of the Speaker, creates an impression that the doll produces the sounds. A discrepancy between sound and lip motion may result in an intermediate percept (this phenomenon is referred to as the *McGurk effect*, McGurk and MacDonald, 1976); for example, when a person listens to "ba" and views a face saying "ga," this combination may induce a perception of "da." This effect is very powerful and can even be seen when the visual and auditory speech signals are not perfectly synchronized (Munhall et al., 1996). On the other hand, skilled musicians do not show the McGurk effect (Proverbio et al., 2016), suggesting that learned attention to sound may overcome the effects of incongruent visual signals.

Extracting Gist-L from spoken Text, however, is not synonymous to identifying all the sounds produced by the Speaker adequately. The most obvious example is listening to a person who speaks a language you do not understand or to a professional using his/her highly specialized jargon. In these examples, the sound may be perceived perfectly well without any chance of extracting Gist-S. On the other hand, misinterpreting sounds is likely to be a major contributor to misinterpretation of the Text with unpredictable consequences for the Text—Gist-L transformation: The misinterpreted Text may belong to the class of T_{IMP} (impossible texts, see Fig. 7.5), leading to projection on an allowed but inadequate IGM and resulting in potentially multiple Gist-L $\neq$ Gist-S.

7.5 Structured variance: the "Chinese Room" problem

The notions of UCM, IPM, and IGM have a certain feature in common: They imply that some directions within the appropriate multidimensional space of elements (orthogonal to the mentioned subspaces) show high stability, whereas other directions (within the subspaces) are expected to be relatively less stable. When a task is repeated multiple times, this leads to a prediction that intertrial variance in directions that span the solution space for specific tasks (those that span the UCM, IPM, or IGM) is expected to be larger than in directions orthogonal to that space. This prediction has been confirmed in many studies of multielement motor actions (reviewed in Latash, 2008) when it is relatively easy to compute the UCM (or, at least, its linear approximation) for a task-related, potentially salient performance variable. Indirect confirmations of this prediction have also been found in studies of force perception (Reschechtko et al., 2018; Cuadra and Latash, 2019). It is intuitively clear that repeating the process of delivering a text message is likely to be associated with variable means of its delivery, i.e., by variability within the IGM, whereas variability orthogonal to the IGM is expected to be low during typical successful communication.

It is important to realize that the prediction of structured intertrial variance does not have to be confirmed for all tasks and with respect to all potentially important variables. This prediction is valid only for salient variables that the central nervous system tries to stabilize, possibly at the expense of destabilization of other variables. In other words, the prediction allows asking the central nervous system questions: "Do you care about this particular variable?", "Do you care about this variable more than about a different variable?" and get a quantitative answer. In these questions, the operational synonym of "care" is "stabilization."

One of the very first studies using the UCM concept (Latash et al., 2001) explored an accurate force production task: The subjects sat in front of a monitor and tried to press naturally with the four fingers of a hand (in other trials, the task was performed by two or three fingers) trying to draw a sine wave oscillating between two targets at a comfortable frequency defined by the auditory metronome. The cursor on the screen showed the current total pressing force produced by all the fingers. Despite the explicit instruction and feedback, the subjects showed an expected structure of intercycle variance ($V_{UCM} > V_{ORT}$) in the space of commands to fingers (so-called, finger *modes*, Danion et al. 2003) only within a relatively small phase range centered about the maximal total force values. In other phases, total force was not stabilized, and variance per dimension was about the same within the UCM and orthogonal to the UCM. So, it seemed that the central nervous system cared about producing accurate magnitudes of total force when it was relatively high and did not care about it when it was low—a very strange conclusion!

When the authors reanalyzed the same data with respect to another mechanical variable produced by the finger forces, the total moment of force produced by the fingers in pronation–supination, the results showed the expected structure of variance, $V_{UCM} > V_{ORT}$, in all phases. This happened despite the fact that the subjects were not told anything about the total moment and were given no visual feedback on that variable. The conclusion drawn by the authors was that the subjects' central nervous systems, based on everyday tasks, stabilized total moment of force preferentially without any special instruction or feedback. Indeed, many of the everyday tasks require accurate rotational action of the hand, while the force production component does not have to be very accurate. For example, when one takes a sip from the glass, the gripping force should only be between the crushing and dropping thresholds, a rather wide range. In contrast, even small errors in the moment of force can lead to spilling the glass contents. This is also true during such actions as writing with the pen, eating with the spoon, using a screwdriver, etc. Note that stabilization of total moment of force favors positive covariation of finger forces on the opposite sides of the pivot, whereas total force stabilization favors negative finger force covariation. This competition (obvious in two-finger tasks) likely resulted in the unexpected lack of total force stabilization.

Although the previous text suggests that the prediction of structured variance is valid for communication, compared with analysis of movements, the situation is more complex because the original space of elements is not well defined. Nevertheless, this prediction, even in its most general form, may be useful to address some of the problems in language-based communication and even philosophy.

Consider, for example, translation from one language to another. This process may be viewed as a sequence Text-L1–Gist–Text-L2, where L1 an L2 stand for the two languages. This example is similar to the ones illustrated in Figs. 7.2 and 7.3, but the order of the two transformations is switched: The first one is illustrated in Fig. 7.3, and the second one—in Fig. 7.2. There is abundance of solutions at each of the two stages, which may be expected to lead to structured variance across repetitive attempts to translate the same text. Indeed, many pieces of classics have been translated multiple times by different translators leading to quite dissimilar Texts, which are all, however, trying to reflect the Gist of the original Text-L1 using different vocabulary and syntax. The fact that Gist-Text transformation allows numerous solutions has been acknowledged for a long time (e.g., in Meaning-Text models, see Mel'cuk 1981) but, to my knowledge, has not been linked to issues of Gist stability and structured variance.

The definition of living system offered in the opening part of the book and the concept of structured variance have implications for issues that belong to philosophy. For example, many years ago, Searle (1992) formulated the famous Chinese Room Problem. There is a room with a small window. You can submit any texts in English and receive translations in Chinese. Your task is to define whether there is a bilingual person in the room with very good knowledge of both English and Chinese or there is a person who does not know Chinese but has access to a multivolume compilation of all possible phrases in English with their perfect translations into Chinese. (Alternatively, the room could have a volume with a fixed set of rules that generate unique, perfect translations into Chinese of all possible English phrases.) There is a straightforward solution for this problem based on the assumed biological specificity of Text-L1−Gist−Text-L2 transformations leading to structured variance: Submit a sequence of 10,000 requests to translate rather complex phrases and repeat one (or more) of the phrases 100 times. The multivolume compilation is expected to produce identical 100 translations for the repeated phrase. A bilingual person is expected to generate varying translations that all preserve the Gist of the original phrase, i.e., produce different solutions that belong to the same IGM.

7.6 Playing chess

The author is not a professional chess player but played chess as a youngster at a decent amateur level, has some level of introspection associated with playing chess, and feels relatively comfortable discussing this game. Playing chess (as well as some other games, such as Go) was for many years considered as a typically human activity where computers would be at a strong disadvantage. However, starting about 20−30 years ago, computers have become at least equal to the best players including World Champions. Note, however, that this was achieved not by understanding how the minds of great players work during chess games and then emulating these processes in computers (cf. Balasubramaniam and Feldman, 2001) but by "brute force": Computers beat humans primarily because of their computational speed and the resulting ability to analyze any position on the chessboard deeper than even the greatest champions. I dare make a prediction: In not-so-distant future, a super-duper computer (similar to the Deep Thought in the classic by Douglas Adams) will be able to analyze the game of chess starting from the initial position all the way down to a draw or win by the whites, and the great game will be all but dead. To me, the fact that the standard Chess program in my Macintosh gives me no chances of winning has been sufficient reason to stop playing chess. It is small consolation that robots are still amazingly clumsy in football (the version called "soccer" in the United States); so, watching games of the Premier League and World Cup remains fun.

Les us, however, forget about computers and consider how natural biological objects (players) analyze the position on the board and come up with a specific move. This is done by thinking helped by memory. Thinking can be defined as the process of getting to a desired mental state from an initial state with tools that represent logical and illogical operations (induction, deduction, intuition, and other processes that are definitely BMC). Mental states take place in spaces of notions (another BMC), which may be viewed as a combination of a multidimensional information space and analysis space. More information affords more variable analyses, which may lead to selecting more successful course of action if the person is not time constrained and to questionable decisions if time is limited.

Many years ago, the author attended seminars in the Institute for Problems of Information Transmission in Moscow. During one of those legendary seminars, one of the participants (a well-known troublemaker) screamed at the presenter: "You speak about information! But do you know what information is?!" Before the presenter could open his mouth, one of the members of the audience replied at a close-to-zero time delay: "Information is something that has problems with transmission explored by our Institute." I have not heard a better definition. However, for the purposes of this discourse, the word "information" is used as a shortcut for *formally quantified salient variables reflecting natural processes in the body*. For example, if someone tells you to produce a force of 20 N, this is information expressed in units of a physical variable. If you are asked to press as strongly as you can, this is information expressed in body-related units. Many tasks do not allow expressing information in physical units, playing chess being one of those, but I am going to assume that there is no qualitative difference between tasks expressed in physical units and those expressed less formally, e.g., in terms of closeness to success in reaching a goal.

This informal definition implies that biologically meaningful information is a reflection of physical (physiological) processes both within the body and at the level of body–environment interactions. Of course, it differs from the formally defined information as in the information theory. Biologically meaningful information is subjective and closely related to the notion of saliency. Many years ago, in the 1960s, my father, Lev Latash, wrote an article for a popular science journal, "Science and Life" (in Russian), with the title "What color was Tal's necktie?". He referred to one of the greatest Grand Masters, World Champion, Mikhail Tal, who took part in the Championship of the USSR that year. In the article, my father wrote that a few days earlier he had visited the Championship site and had definitely seen Tal but could not remember the color of Tal's necktie although the necktie had been in his clear view. This was used as an example of visual information filtering into salient and nonsalient components, and only the salient ones were retained in memory. Along similar lines, here and later in this part, I imply only salient information.

Thought can be viewed as the process of creating low-dimensional outcomes from very high-dimensional sets of variables (cf. Rabinovich et al., 2015). Stability of the outcomes of thought processes (their robustness) depends on many unknown factors including exactness of those processes. For example, a mathematical theorem proven once is robust because it represents the result of a sequence of exact logical steps. In everyday life, logical steps are typically not exact and, as a result, stability drops with the number of steps in the solution process (ambiguity accumulates). In such a situation, ensuring stability of a solution can be done by repeating the process of solving a mental problem along different routes and coming to the same, or sufficiently similar, outcomes. In other words, stability can be tested by perturbing the thought process and observing divergence of results or convergence to the same result. This conclusion is similar to the one illustrated in Fig. 4.2, which justifies intertrial analysis of variance as a method of quantifying stability of action.

For motor action and kinesthetic perception, the information space may be viewed as analogous to the space of sensory signals (the afferent process) and memory, whereas the analysis or action space is analogous to the efferent process. Of course, this analogy is imprecise, for example, because the two processes are nonindependent due to the action-perception coupling. The obvious link is the system of gamma-motoneurons, which are coactivated with alpha-motoneurons (the output of the efferent process) and affect sensitivity of spindle sensory endings to muscle length and velocity (a major input of the afferent process). This seems to be true also for mental processes: analysis of a problem may by itself lead to modification of the problem.

For mental processes, such as analysis of a position on the chessboard, information space is related to the current position on the board, memory-stored similar positions from one's own experience or literature, and, possibly, to the apparent state of the opponent (e.g., whether he or she is tired or nervous). For example, according to the lore, the great World Champion of the past, Emanuel Lasker (1868-1941), who was a psychologist and a mathematician (see the Lasker−Noether theorem!), tried to select moves that would not be necessarily objectively optimal given the position on the board, but rather most uncomfortable for the opponent. The action space is related to analysis of the position and progress in solving the task of selecting an optimal next move. Success or failure of action can be described with such characteristics as depth of analysis and its exactness. The last term combines estimated change in the probability of winning the game, stability of the positive outcome (increased probability of winning), and/or instability on the negative outcome (decreased probability of losing) given the unpredictable moves by the opponent.

Relations between the information (INF) and action (ACT) spaces are illustrated in Fig. 7.6. Although both spaces are multidimensional, I illustrate them with single axes (as in Fig. 7.2). The family of curves starting from different points on the INF axis

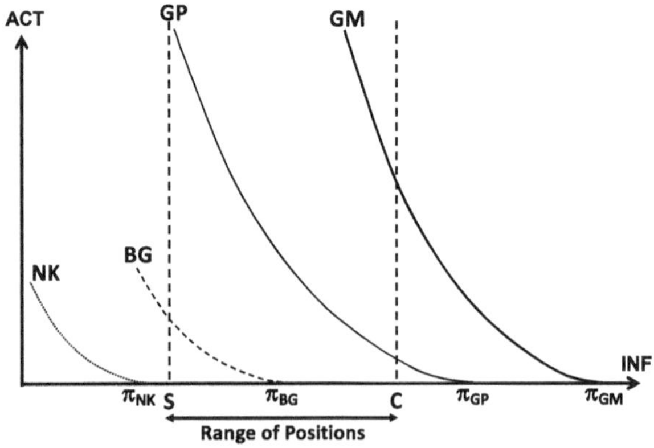

Figure 7.6 Relations between the information (INF) and action (ACT) spaces for the game of chess are shown for four hypothetical players differing in the threshold of the ACT(INF) dependence: π_{NK} (no knowledge of the game, NK *curve*), π_{BG} (beginner, BG *curve*), π_{GP} (good player, GP *curve*), and π_{GM} (Grand Master, GM *curve*). The range of possible positions on the board is shown as S (simple) to C (complex).

reflects relations between available information and depth and accuracy of analysis, i.e. action. Each curve is defined by its threshold, a parameter defining the dependence of action on available information. This makes the scheme in Fig. 7.6 another specific example of parametric control.

If one does not know the rules of chess, the ACT(INF) curve starts far to the left (the NK curve in Fig. 7.6), beyond the range of possible positions, from S (simple) to C (complex). A beginner (the BG curve in Fig. 7.6) may be able to find acceptable solutions for relatively simple positions on the board but not for more complex positions. When you become a better player, you become able to find optimal moves in more and more complex positions: From winning with an extra queen or rook to winning with an extra pawn and checkmating with a bishop and a knight. This is reflected in a shift of the ACT(INF) curve to the right (the thin curve labeled GP). Ultimately, a Grand Master can find optimal moves even for very complex positions as illustrated with the GM curve.

Formally, one's knowledge of the game of chess and experience, combined with other factors, such as ability to concentrate, define threshold for the ACT(INF) curve. The threshold (π) reflects the most complex position on the board for which the player is able to find a move with positive values of action, i.e. increasing one's chances to win or decreasing chances of a loss. This parameter reflects one's initial abilities to analyze the situation on the board as illustrated by the different thresholds for the mentioned players, π_{NK} (no knowledge of the game), π_{BG} (beginner), π_{GP} (good player), and π_{GM} (Grand Master). During the game, a player can concentrate on analysis of the position and thus shift the parameter to the right or relax and allow it to slip to the left.

Analysis of a position on the chessboard, as well as any thought process, can start with a set of initial (current) parameters and state variables and/or with a set of desired values of state variables, e.g., a desired position on the chessboard corresponding to higher chances of winning the game. Thinking without sufficient relevant information is not going to be successful (beyond the left flank in Fig. 7.6); this can happen, for example, during games played without looking at the board when one of the players forgets the current configuration of the pieces. Using one's memory and concentrating, and spending more time on analysis of the position, leads to a change in π toward higher values, translates the ACT(INF) curve to the right, and improves chances of finding an optimal move.

7.6.1 Information–action coupling

Fig. 7.6 suggests that information and action spaces are coupled and that this coupling can be modified by setting specific parameters (such as π in Fig. 7.6). The coupling is bidirectional because information leads to action, which in turn changes information; for example, a move on the chessboard, actual or imagined, changes the position. The notion of information–action coupling is related to the classical notion of perception–action coupling (reviewed in Kugler and Turvey, 1987; Richardson et al., 2008; Whitall and Clark, 2018) as introduced within the ecological approach to psychology pioneered by Gibson (1979). Indeed, perception–action coupling may be seen as special case of information–action coupling.

Coupling between afferent (sensory) and efferent (motor) signals has been known for a long time. Indeed, the classical stretch reflex (Liddell and Sherrington, 1924) can be seen as an example of perception–action coupling; the same can be said about many other reflexes and reflex-like reactions to sensory signals, such as preprogrammed reactions (Phillips, 1969; Marsden et al., 1979; Forssberg, 1979; Prochazka et al., 2002), also addressed as triggered reactions, long-loop reflexes, and M_{2-3}.

Coupling between sensory and motor signals mediated by the stretch reflex has been discussed earlier (Sections 2.2.1 and 5.1). In this example, sensory signals reflect muscle length (and possibly other mechanical variables), whereas motor signals result in muscle activation and force. Specific shape of this coupling, e.g., the shape of the dependence between active muscle force on muscle length, can be controlled by setting a parameter, λ, associated with the threshold of the reflex and also with subthreshold depolarization of the corresponding alpha motoneuronal pool (reviewed in Feldman, 2015). Note that a change in muscle force leads to a change in muscle length defined by the peripheral mechanics. As a result, setting a value of λ does not define values of either of the two variables; instead, it defines an equilibrium point for the system "muscle + reflex + external force field."

There are more complex examples of perception–action coupling, e.g., in tasks involving motion of pairs of effectors reflected in unintentional changes in the relative phase of the effector trajectories, e.g., with a change in movement frequency or with a change in the mechanical properties of one of the effectors (reviewed in Kelso, 1995; Sternad, 1998). Typically, in such tasks, only the in-phase and out-of-phase movements are stable, and the in-phase one shows higher stability (see Section 5.4.2). Even in relatively simple two-effector actions, instructions (information) could qualitatively change stability properties of specific patterns. Consider, for example, a pattern involving rhythmical motion of the wrist and elbow joints within an arm. The flexion–flexion and extension–extension patterns (in-phase) were shown to be more stable than the flexion–extension pattern (out-of-phase). This was reflected, in particular, in an unintentional switch from the out-of-phase pattern to the in-phase one with an increase in the movement frequency. Consider, however, that you tell the subject: "Imagine that you are a waiter with the tray on the hand; now move the tray away from and toward the body quickly" (an example suggested by Peter Greene). The instructed motion involves the flexion–extension pattern (out-of-phase), which is supposedly less stable, but, under this instruction, speeding the movement up does not lead to a switch to the in-phase pattern.

Another example of a change in the relative stability of in-phase and out-of-phase patterns was provided in studies of two-hand motion under modified visual feedback (Hommel et al., 2001; Mechsner et al., 2001). When visual feedback was reversed with the help of a simple mechanical contraption to make in-phase movements look out-of-phase and vice versa, the previously more stable in-phase pattern became less stable as compared with the out-of-phase pattern. Note that the natural, somatosensory feedback was unchanged! These examples suggest that information variables, whether presented as modified visual feedback or as a verbal instruction, are able to change stability properties of patterns defined by perception–action coupling.

Stability of relative phase at a specific value is a trademark of two-effector actions. This is true for tasks involving two effectors within a single person and for tasks performed by two persons watching each other as it has been shown in experiments when the subjects were asked to swing pendulums held in the dominant hands while watching each other (Schmidt et al., 1990; Schmidt and Turvey, 1994). Under such conditions, the subjects naturally synchronized the trajectories of the two pendulums and showed all the trademark features of changes in the movement kinematics and relative phase with a change in the movement frequency. In this case, the sensory variable responsible for the coupling was visual, but it could probably be replaced with purely informational variable, for example, voice describing motion of the other person's pendulum: "front-back-front-back, etc."

The idea of perception–action coupling has been developed in contrast to the dominant view in cognitive psychology that sensory signals are used by the brain to update the intrinsic picture of the world and then to use this picture for the planning and generation of motor actions. The idea of coupling implies that a stream of sensory signals can lead to adjustments of actions without updating the picture of the world. Can action–perception coupling be mediated by intrinsic neural processes such that actions are coupled to intrinsic sources of information, e.g., those stored in memory? In other words, can one imagine something and, by doing that, produce an unintentional adjustment of an ongoing action without realizing that the adjustment has happened? Introspection suggests that this is likely true, but the author is unaware of properly controlled experimental studies demonstrating such effects.

Getting back to playing chess, the closest example to perception–action coupling is playing chess under very short time control (e.g., so-called 3-min blitz games). During such games, there is no time for any kind of serious update of the position on the chessboard and analysis. As most players know, moves are nearly "automatic," of a stimulus-response type, and, as a result, are commonly suboptimal.

One of the common topics of discussions with my colleagues is whether the brain is a "complex computer." Indeed, in some situations, including those studied in the sterile environment of a typical laboratory, the brain seems to produce behavioral patterns that are linked in a rather straightforward way to the task. In other words, the central nervous system seems to make decisions based on a relatively simple set of rules, similar to the Turing machine. As soon as the situation becomes less sterile and more typical of functional everyday tasks, the analogy fails: Behavioral patterns become much more complex and less predictable, as one could expect from a time-varying physical system. They, however, retain an important feature: Stability of salient performance variables produced by much more variable patterns in a high-dimensional system of EVs. Examples from the fields of motor actions and perception have been discussed in earlier parts of the book. They are likely to be produced by neuronal organizations that share typical features across important behaviors, action, perception, and—as claimed in this Part—cognition.

This approach allows to define *cognition* as *an ability to organize stable subspaces in high-dimensional spaces of mental variables* (e.g., notions), likely with participation of action- and perception-related EVs. Along similar lines, one can define *intelligence* as *an ability to select such stable subspaces and to move from one stable subspace to another subspace in a way that maximizes reward* for the actor. This definition makes intelligence directly linked to the space of EVs where relevant processes take place.

Many years ago, I was confronted with a review of one of my manuscripts claiming that, in that paper, I assumed intelligence of the object of study and of inputs into that object. This was a tough criticism to handle because intelligence was not defined. So, I asked a colleague, a professor of philosophy in our Department, what intelligence was. He laughed, said "welcome to the community of philosophers," and explained to me that there was no single definition for intelligence, and there were many types of intelligence including, in particular, "movement intelligence."

At the time, the explanation was not very helpful, but now I start to understand that indeed different types of intelligence exist corresponding to different spaces of EVs where the behaviors, intelligent or not, take place. There are indeed "motor intelligence" "sensory intelligence" "chess intelligence" "common sense intelligence" and many more. Maybe different types of intelligence covary across the population, but such covariation seems to be not very strong given the large numbers of clumsy smart persons and excellent athletes who are struggling with their homework.

What about this book? Was it written in an intelligent way? Does it create stable subspaces in the minds of the readers that match the author's ideas and goals? Only readers may be able to answer these questions, and I am not even sure about that.

Problems and conjectures

8

"Computers are not intelligent.
They only think they are."
Osburn's Axiom (Bloch, 2003)

Problem 1: Is it easier or more complicated for the brain to control a system within the body with many elements?

Conjecture 1: Control with setting global parameters benefits from more elements, which allow combining flexibility of performance (e.g., performing secondary tasks) with its stability.

Explanation and Examples: More elements frequently make the system's behavior more predictable. In physics, rigid bodies behave much more predictably than individual elemental particles within those bodies. To use a more biological example, it is much easier for a politician to convince a crowd than an individual. Multiple elements lead to high-dimensional solution spaces for typical tasks; within those spaces, other tasks can be performed while keeping performance of the primary tasks stable.

Problem 2: Why did evolution come up with parametric control while engineers prefer to build robots that produce actions with precomputed forces?

Conjecture 2: The evolution came up with the most advantageous scheme ensuring survival and success given the unpredictable environment. Engineers build objects that have a limited range of goals; success in natural selection is not one of those goals.

Explanation and Examples: If *homo sapiens* and other higher animals were exterminated (e.g., by a self-created ecological catastrophe or another World War), all man-made machines, including the most sophisticated robots, would rust and fall apart relatively quickly, while viruses, bacteria, and similar organisms would start a new round of evolution, based on parametric control, possibly leading to a more successful outcome.

Problem 3: How detailed is the neural control of action at the highest (task-related) level?

Conjecture 3: During typical actions, control is low dimensional corresponding to parameters (e.g., referent coordinates) for a handful of salient variables only.

Explanation and Examples: Numerous studies reviewed in this book have shown that consistent performance at the task level is frequently accompanied by highly variable performance at the level of elements. Hence, time profiles of element-level referent coordinates are likely not prescribed but emerge given the unpredictable, nonrepetitive external forces, sensory signals, and intrinsic states of the body.

Physics of Biological Action and Perception. https://doi.org/10.1016/B978-0-12-819284-9.00008-5

Problem 4: Can performance variables be stabilized at several levels of a control hierarchy?

Conjecture 4: Yes, but this requires a nontrivial combination of stability and optimality at the highest level.

Explanation and Examples: There is an inherent trade-off between stability at two levels of a hierarchy because "good variability" (within the uncontrolled manifold, UCM) at the higher level translates into "bad variability" (within the orthogonal to the UCM space, ORT) at the lower level. If, at the higher level, the variance is structured reflecting stability of a salient variable but its overall magnitude is very low (variance both within the UCM and within ORT is small), there is a possibility of having similarly structured variance at the lower level.

Problem 5: What are the neurophysiological mechanisms involved in setting referent coordinates at the task level or at any other level of the control hierarchy?

Conjecture 5: Subthreshold depolarization of the membranes of corresponding (at this time, mostly unknown) neuronal pools.

Explanation and Examples: This method of control makes use of the threshold property of neuronal membranes and turns neurons into adaptive elements, sensitive to other inputs, in particular those that carry relevant sensory information on salient variables. Using a suprathreshold control signal forces neurons to generate action potentials at their highest frequency independently of other inputs, which eliminates their functionally important feature of adaptive behavior.

Problem 6: How can the central nervous system (CNS) minimize costs it cannot *perceive*?

Conjecture 6: Costs are minimized (or, more exactly, they do not deviate from minimal values by too much) as consequences of learned patterns of variables at the control level and laws of nature involved in the generation of actions. The costs are perceived by researches, not by moving animals.

Explanation and Examples: The CNS of an animal does not compute any costs because it is not a computational device but a natural object that behaves according to laws of nature. Preferred solutions create an impression of minimizing a cost when they do not violate that cost by too much. As a result, different cost functions have been suggested to account, with about equal success, for typically observed behavioral patterns.

Problem 7: Why do people and other animals coactivate agonist–antagonist muscle pairs? This seems to be energetically suboptimal.

Conjecture 7: Muscle coactivation is an epiphenomenon related to ensuring stability of action at the level of control. Nonzero activation of opposing muscles reflects availability of at least two variables at the control level per effector (e.g., joint or digit) making the control system abundant and affording controlled stability of salient performance variables.

Explanation and Examples: Changing control variables (λ) to muscles has no effect on action when λ is larger than the actual muscle length. Hence, for any action, if only one muscle (agonist) is active, only one variable at the control level can be used to modify the action. As a result, there is no abundance at the control level and no possibility to stabilize action by covariation of control variables. Coactivating muscles creates a safety zone where small changes in λ preserve the abundance at the control level. Coactivation to increase stability at the level of peripheral mechanics can work only for effectors with fixed origin.

Problem 8: Do drifts in neural variables (and performance) happen at all times or only in cases of sensory (e.g., visual) deprivation?

Conjecture 8: Drifts of neural control variables happen continuously across tasks and conditions representing the natural feature of physical systems to move to states with lower potential energy. If salient feedback is available, visual or proprioceptive, these drifts may be corrected. Otherwise, they lead to accumulation of performance drifts.

Explanation and Examples: Performance drifts are accompanied by drifts in the effector referent coordinate and apparent stiffness reflecting drifts in the referent coordinates for the muscles. Force drift is not accompanied by a drift during force matching by the other hand or in memory-based performance, during force matching by the same hand after a brief rest period.

Problem 9: What is the origin of perception?

Conjecture 9: Perception is a reported (including self-reported) creation of an iso-perceptual manifold (IPM) in the high-dimensional space of afferent and efferent variables.

Explanation and Examples: There are many everyday observations pointing at the existence of an IPM. Without a corresponding IPM, percepts would become unstable and unlikely to be reported as percepts. Kinesthetic perception is typically veridical and shows stability under such actions as cocontraction and isometric force production. This may be seen as motion along the corresponding IPM.

Problem 10: How are multiple sensory signals reduced to a few high-order signals on salient variables?

Conjecture 10: There are synergic organizations in the space of afferent (sensory) and efferent signals stabilizing their functions reflecting the magnitudes of salient physical variables. Such many-to-few transformations happen in both perceptual and motor domains and may also form the basis for stability of outcomes of cognitive processes.

Explanation and Examples: The synergic organization of the IPM is supported by observations of less accurate percepts at the level of elements contributing to a multi-element task. The importance of efferent processes for perception has been shown in many studies including those with muscle vibration and with movements produced by direct electrical stimulation of muscles. Stability of low-dimensional outcomes is common across the motor, perceptual, and cognitive domains.

Problem 11: What is the role of sensory signals in ensuring stable performance?

Conjecture 11: Tasks practiced under specific sensory feedback are likely to develop means of ensuring stability that rely on that specific feedback. Other task components, including implicit ones, are likely to be stabilized using intrinsic circuitry (e.g., back-coupling loops within the CNS).

Explanation and Examples: Tasks practiced under visual feedback show large drifts and loss of stability if feedback is removed. This does not happen with task components that are typically performed only under natural somatosensory feedback (e.g., typically, humans do not drop gripped objects, even when they hold them for a long time).

Problem 12: Are there dedicated physiological structures responsible for stability of action/perception/gist?

Conjecture 12: There are many of those at different levels of the CNS, from the spinal cord to brain structures.

Explanation and Examples: At the spinal level, examples are the system of recurrent inhibition stabilizing the output of an alpha-motoneuronal pool and the stretch reflex stabilizing the equilibrium state of the muscle acting against an external load. At the brain level, loss of action stability is associated with dysfunction of some structures (e.g., nuclei within the basal ganglia, the cerebellum, and the associated pathways), while stability is relatively resistant to injury to other brain structures (e.g., following mild cortical stroke).

Problem 13: What happens with gist stability when speaking a foreign language?

Conjecture 13: If the language is well known to the speaker, gist is stable. This may be viewed as a definition of knowing the language well.

Explanation and Examples: Poor knowledge of a foreign language leads to frequent episodes of misunderstanding during both speech production and comprehension, i.e., loss of gist stability. Another factor contributing to poor understanding may be the lack of an established set of allowed iso-gist manifolds, which requires knowledge not only of the language but also of cultural constraints and customs.

Problem 14: How is stability of multiperson interactions and group behaviors achieved?

Conjecture 14: It may be achieved in a dictatorial way (one person prescribes what all others have to do and issues error correction commands when errors occur) or by uniting the group into a dynamical system (by setting rules of interaction and common goals).

Explanation and Examples: Consider the conductor of a choir. The conductor can either tell each singer what and how to sing or set for them a target value of sound and rely on their ears. In the former case, stability of performance will be poor. In the latter case, performance will be more stable, even in cases of major perturbations (e.g., when one of the singers stops performing altogether).

Problem 15: Why is the research community so unwilling to accept the idea of parametric control and its implementation in the form of control with referent coordinates?

Conjecture 15: The success of engineering in the 20th century and the natural human tendency to look for alreadily available solutions are the main reasons.

Explanation and Examples: It is very tempting to adopt a successful approach from a well-developed field of human activity and apply it to an object of study that belongs to a different field, even if this object is completely beyond one's comprehension (as the human brain is). The success of engineering-based approaches in building moving machines that can simulate successfully some of the salient features of animal (human) behavior has been the source of encouragement for champions of this approach. The fact that this approach borders with creationism and is nonscientific has been ignored.

Problem 16: Where do goals for actions come from?

Conjecture 16: They probably come from interactions of sensory signals of all modalities (including those from intrinsic organs) with memory.

Explanation and Examples: So far, there has been only one approach, the dynamic field theory developed by Gregor Schöner and colleagues (Erlhagen and Schöner, 2002; Schöner and Thelen, 2006), which tries to address decision-making processes, including target selection, based on laws of nature. This theory involves complex, dynamic interactions among signals from different sources, sensory and memory-based, within the brain. In other approaches, goals come from unknown processes (beyond my comprehension, BMC).

Problem 17: What variables are encoded in characteristics of activation of brain *neuronal* populations?

Conjecture 17: Neurons in the primary motor cortex encode time profiles of referent coordinate for the task-specific effector. Neurons in subcortical structures encode the mapping from the task-level referent coordinates (RC) to effector-level RCs that defines action stability.

Explanation and Examples: The first statement is compatible with a body of literature on neuronal population vectors and their links to performance variables, as discussed in much detail in a recent review by Feldman (2019). The second statement is supported by observations of impaired control of action stability in patients with subcortical brain disorders, particularly those involving the cerebellum and basal ganglia, not as pronounced after cortical stroke.

Problem 18: What is cognition? What is intelligence?

Conjecture 18: Cognition is an ability to stabilize subspaces in high-dimensional spaces of elemental mental, motor, and sensory variables. Intelligence is an ability to move consistently from one such subspace to another and to select such trajectories in a way that gives advantage, actual or imaginary, to the animal.

Explanation and Examples: Functional motor and sensory states are stable; unstable states are useless with one possible exception of intermediate states observed during transitions between stable states. Similar to stable motor and sensory spaces, stable mental states are embedded in spaces of imagined (e.g., extracted from memory) physical variables combined with actual motor and sensory processes and corresponding elemental variables within the body. Intelligence may be seen as a problem-solving ability to find stable solutions that give the largest reward to the actor. It is task and space specific: A chess grand master may be clumsy, i.e., shows low intelligence in motor tasks, whereas the best basketball player may be hopeless in chess and similar mental games.

References

Abbs, J.H., Gracco, V.L., 1984. Control of complex motor gestures: orofacial muscle responses to load perturbations of the lip during speech. Journal of Neurophysiology 51, 705–723.

Akulin, V.M., Carlier, F., Solnik, S., Latash, M.L., 2019. Sloppy, but acceptable, control of biological movement: algorithm-based stabilization of subspaces in abundant spaces. Journal of Human Kinetics 67, 49–72.

Alexander, M.R., 2002. Energetics and optimization of human walking and running: the 2000 Raymond Pearl memorial lecture. American Journal of Human Biology 14, 641–648.

Alexandrov, A.V., Frolov, A.A., Massion, J., 2001. Biomechanical analysis of movement strategies in human forward trunk bending. I. Modeling. Biological Cybernetics 84, 425–434.

Alexandrov, A.V., Lippi, V., Mergner, T., Frolov, A.A., Hettich, G., Husek, D., 2017. Human-inspired eigenmovement concept provides coupling-free sensorimotor control in humanoid robot. Frontiers in Neurorobotics 11, 22.

Ambike, S., Mattos, D., Zatsiorsky, V.M., Latash, M.L., 2016a. Synergies in the space of control variables within the equilibrium-point hypothesis. Neuroscience 315, 150–161.

Ambike, S., Mattos, D., Zatsiorsky, V.M., Latash, M.L., 2016b. Unsteady steady-states: central causes of unintentional force drift. Experimental Brain Research 234, 3597–3611.

Ambike, S., Mattos, D., Zatsiorsky, V.M., Latash, M.L., 2016c. The nature of constant and cyclic force production: unintentional force-drift characteristics. Experimental Brain Research 234, 197–208.

Ambike, S., Mattos, D., Zatsiorsky, V.M., Latash, M.L., 2018. Systematic, unintended drifts in the cyclic force produced with the fingertips. Motor Control 22, 82–99.

Ambike, S., Paclet, F., Zatsiorsky, V.M., Latash, M.L., 2014. Factors affecting grip force: anatomy, mechanics, and referent configurations. Experimental Brain Research 232, 1219–1231.

Ambike, S., Zatsiorsky, V.M., Latash, M.L., 2015a. Processes underlying unintentional finger force changes in the absence of visual feedback. Experimental Brain Research 233, 711–721.

Ambike, S., Zhou, T., Zatsiorsky, V.M., Latash, M.L., 2015b. Moving a hand-held object: reconstruction of referent coordinate and apparent stiffness trajectories. Neuroscience 298, 336–356.

Aoyama, T., Kaneko, F., 2011. The effect of motor imagery on gain modulation of the spinal reflex. Brain Research 1372, 41–48.

Arbib, M.A., Iberall, T., Lyons, D., 1985. Coordinated control programs for movements of the hand. In: Goodwin, A.W., Darian-Smith, I. (Eds.), Hand Function and the Neocortex. Springer Verlag, Berlin, pp. 111–129.

Arce, F., Novick, I., Mandelblat-Cerf, Y., Vaadia, E., 2010. Neuronal correlates of memory formation in motor cortex after adaptation to force field. Journal of Neuroscience 30, 9189–9198.

Arias, P., Espinosa, N., Robles-García, V., Cao, R., Cudeiro, J., 2012. Antagonist muscle co-activation during straight walking and its relation to kinema1tics: insight from young, elderly and Parkinson's disease. Brain Research 1455, 124–131.

Aruin, A.S., 2003. The effect of changes in the body configuration on anticipatory postural adjustments. Motor Control 7, 264–277.

Aruin, A.S., Almeida, G.L., Latash, M.L., 1996. Organization of a simple two-joint synergy in individuals with Down syndrome. American Journal of Mental Retardation 101, 256–268.

Asaka, T., Wang, Y., Fukushima, J., Latash, M.L., 2008. Learning effects on muscle modes and multi-mode synergies. Experimental Brain Research 184, 323–338.

Asaka, T., Yahata, K., Mani, H., Wang, Y., 2011. Modulations of muscle modes in automatic postural responses induced by external surface translations. Journal of Motor Behavior 43, 165–172.

Asanuma, H., 1973. Cerebral cortical control of movements. Physiologist 16, 143–166.

Asatryan, D.G., Feldman, A.G., 1965. Functional tuning of the nervous system with control of movements or maintenance of a steady posture. I. Mechanographic analysis of the work of the limb on execution of a postural task. Biophysics 10, 925–935.

Babinski, F., 1899. De l'asynergie cerebelleuse. Revue Neurologique 7, 806–816.

Balasubramaniam, R., Feldman, A.G., 2001. Some robotic imitations of biological movements can be counterproductive. Behavioral and Brain Sciences 24, 1050–1051.

Belen'kii, V.Y., Gurfinkel, V.S., Pal'tsev, Y.I., 1967. Elements of control of voluntary movements. Biofizika 10, 135–141.

Bennett, D.J., 1993. Electromyographic responses to constant position errors imposed during voluntary elbow joint movement in human. Experimental Brain Research 95, 499–508.

Bennett, D.J., Hollerbach, J.M., Xu, Y., Hunter, I.W., 1992. Time-varying stiffness of human elbow joint during cyclic voluntary movement. Experimental Brain Research 88, 433–442.

Berger, W., Trippel, M., Discher, M., Dietz, V., 1992. Influence of subjects' height on the stabilization of posture. Acta Otolaryngology 112, 22–30.

Berkinblit, M.B., Feldman, A.G., Fukson, O.I., 1986. Adaptability of innate motor patterns and motor control mechanisms. Behavioral and Brain Sciences 9, 585–638.

Bernstein, N.A., 1966. Essays on the Physiology of Movements and Physiology of Activity. Meditsina, Moscow (in Russian).

Bernstein, N.A., 1930. A new method of mirror cyclographie and its application towards the study of labor movements during work on a workbench. Hygiene, Safety and Pathology of Labor. # 5, pp. 3–9, and # 6, pp. 3–11 (in Russian).

Bernstein, N.A., 1935. The problem of interrelation between coordination and localization. Archives of Biological Sciences 38, 1–35 (in Russian).

Bernstein, N.A., 1947. On the Construction of Movements. Medgiz, Moscow (in Russian).

Bernstein, N.A., 1967. The Co-ordination and Regulation of Movements. Pergamon Press, Oxford.

Bernstein, N.A., 1996. On dexterity and its development. In: Latash, M.L., Turvey, M.T. (Eds.), Dexterity and its Development. Erlbaum Publ., Mahwah, NJ, pp. 1–244.

Bernstein, N.A., 2003. Contemporary Studies on the Physiology of the Neural Process. Smysl, Moscow, Russia (in Russian).

Bicchi, A., Gabiccini, M., Santello, M., 2011. Modelling natural and artificial hands with synergies. Philosophical Transactions of the Royal Society of London B Biological Sciences 366, 3153–3161.

Birbaumer, N., Piccione, F., Silvoni, S., Wildgruber, M., 2012. Ideomotor silence: the case of complete paralysis and brain-computer interfaces (BCI). Psychological Research 76, 183–191.

Birglen, L., Laliberté, T., Gosselin, C.M., 2008. Underactuated Robotic Hands. Springer Tracts in Advanced Robotics). Springer, NY.

Bizzi, E., Accornero, N., Chapple, W., Hogan, N., 1982. Arm trajectory formation in monkeys. Experimental Brain Research 46, 139–143.

Bizzi, E., Polit, A., Morasso, P., 1976. Mechanisms underlying achievement of final head position. Journal of Neurophysiology 39, 435–444.

Blasko, D.G., Hall, M.D., 1998. Influence of prosodic boundaries on comprehension of spoken english sentences. Perceptual and Motor Skills 87, 3–18.

Bloch, A., 2003. Murphy's Law. The 26th Anniversary Edition. Berkeley Publishing Group, NY.

Bolbecker, A.R., Apthorp, D., Martin, A.S., Tahayori, B., Moravec, L., Gomez, K.L., O'Donnell, B.F., Newman, S.D., Hetrick, W.P., 2018. Disturbances of postural sway components in cannabis users. Drug and Alcohol Dependence 190, 54–61.

Bongaardt, R., 2001. How bernstein conquered movement. In: Latash, M.L., Zatsiorsky, V.M. (Eds.), Classics in Movement Science. Human Kinetics, Urbana, IL, pp. 59–84.

Booth, F.W., Weeden, S.H., Tseng, B.S., 1994. Effect of aging on human skeletal muscle and motor function. Medicine and Science in Sports and Exercise 26, 556–560.

Borg, G.A., 1982. Psychophysical bases of perceived exertion. Medicine and Science in Sports and Exercise 14, 377–381.

Bosco, G., Poppele, R.E., 2002. Encoding of hindlimb kinematics by spinocerebellar circuitry. Archives Italiennes de Biologie 140, 185–192.

Bottasso, C.L., Prilutsky, B.I., Croce, A., Imberti, E., Sartirana, S., 2006. A numerical procedure for inferring from experimental data the optimization cost functions using a multibody model of the neuro-musculoskeletal system. Multibody System Dynamics 16, 123–154.

Bove, M., Courtine, G., Schieppati, M., 2002. Neck muscle vibration and spatial orientation during stepping in place in humans. Journal of Neurophysiology 88, 2232–2241.

Bowler, R.M., Nakagawa, S., Drezgic, M., Roels, H.A., Park, R.M., Diamond, E., Mergler, D., Bouchard, M., Bowler, R.P., Koller, W., 2007. Sequelae of fume exposure in confined space welding: a neurological and neuropsychological case series. Neurotoxicology 28, 298–311.

Brandon, R.N., 1990. Adaptation and Environment. Princeton Univ. Press, Princeton, NJ.

Brandon, R.N., 1997. Does biology have laws? The experimental evidence. Philosophy of Science 64, S444–S457. Proceedings.

Brink, E.E., Suzuki, I., Timerick, S.J.B., Wilson, V.J., 1984. Directional sensitivity of neurons in the lumbar spinal cord to neck rotation. Brain Research 323, 172–175.

Brooks, S.V., Faulkner, J.A., 1994. Skeletal muscle weakness in old age: underlying mechanisms. Medicine and Science in Sports and Exercise 26, 432–439.

Brown, M.C., Engberg, I., Matthews, P.B., 1967. The relative sensitivity to vibration of muscle receptors of the cat. Journal of Physiology 192, 773–800.

Bruton, M., O'Dwyer, N., 2018. Synergies in coordination: a comprehensive overview of neural, computational, and behavioral approaches. Journal of Neurophysiology 120, 2761–2774.

Buckingham, G., 2014. Getting a grip on heaviness perception: a review of weight illusions and their probable causes. Experimental Brain Research 232, 1623–1629.

Burgess, P.R., Clark, F.J., 1969. Characteristics of knee joint receptors in the cat. Journal of Physiology 203, 317–335.

Burrack, A., Brugger, P., 2005. Individual differences in susceptibility to experimentally induced phantom sensations. Body Image 2, 307–313.

Cabel, D.W., Cisek, P., Scott, S.H., 2001. Neural activity in primary motor cortex related to mechanical loads applied to the shoulder and elbow during a postural task. Journal of Neurophysiology 86, 2102–2108.
Cafarelli, E., Kostka, C.E., 1981. Effect of vibration on static force sensation in man. Experimental Neurology 74, 331–340.
Canales, J., 2015. The Physicist and the Philosopher: Einstein, Bergson, and the Debate That Changed Our Understanding of Time. Princeton University Press, Princeton, NJ.
Cannon, S.C., Zahalak, G.I., 1982. The mechanical behavior of active human skeletal muscle in small oscillations. Journal of Biomechanics 15, 111–121.
Celnik, P., 2005. Understanding and modulating motor learning with cerebellar stimulation. The Cerebellum 14, 171–174.
Cheng, M., Kato, M., Tseng, C.H., 2017. Gender and autistic traits modulate implicit motor synchrony. PLoS One 12 (9), e0184083.
Clark, F.J., Burgess, P.R., 1975. Slowly adapting receptors in cat knee joint: can they signal joint angle? Journal of Neurophysiology 38, 1448–1463.
Clifford, C.W., Mareschal, I., Otsuka, Y., Watson, T.L., 2015. A Bayesian approach to person perception. Consciousness and Cognition 36, 406–413.
Cole, K.J., Abbs, J.H., 1987. Kinematic and electromyographic responses to perturbation of a rapid grasp. Journal of Neurophysiology 57, 1498–1510.
Collins, J.S., Robinson, E.J., 2005. Can one written word mean many things? Prereaders' assumptions about the stability of written words' meanings. Journal of Experimental Child Psychology 90, 1–20.
Coombes, S.A., Corcos, D.M., Vaillancourt, D.E., 2011. Spatiotemporal tuning of brain activity and force performance. Neuroimage 54, 2226–2236.
Corcos, D.M., Agarwal, G.C., Flaherty, B.P., Gottlieb, G.L., 1990. Organizing principles for single-joint movements: IV. Implications for isometric contractions. Journal of Neurophysiology 64, 1033–1042.
Corcos, D.M., Gottlieb, G.L., Agarwal, G.C., 1989. Organizing principles for single joint movements. II. A speed-sensitive strategy. Journal of Neurophysiology 62, 358–368.
Corcos, D.M., Gottlieb, G.L., Penn, R.D., Myklebust, B., Agarwal, G.C., 1986. Movement deficits caused by hyperexcitable stretch reflexes in spastic humans. Brain 109, 1043–1058.
Cordo, P., Carlton, L., Bevan, L., Carlton, M., Kerr, G.K., 1994. Proprioceptive coordination of movement sequences: role of velocity and position information. Journal of Neurophysiology 71, 1848–1861.
Craske, B., 1977. Perception of impossible limb positions induced by tendon vibration. Science 196, 71–73.
Cross, M.J., McCloskey, D.I., 1973. Position sense following surgical removal of joints in man. Brain Research 55, 443–445.
Crowninshield, R.D., Brand, R.A., 1981. A physiologically based criterion of muscle force prediction in locomotion. Journal of Biomechanics 14, 793–801.
Cruse, H., Bruwer, M., 1987. The human arm as a redundant manipulator: the control of path and joint angles. Biological Cybernetics 57, 137–144.
Cruse, H., Wischmeyer, E., Bruwer, M., Brockfield, P., Dress, A., 1990. On the cost functions for the control of the human arm movement. Biological Cybernetics 62, 519–528.
Crutcher, M.D., DeLong, M.R., 1984. Single cell studies of the primate putamen, parts 1 and 2. Experimental Brain Research 53, 233–258.
Cuadra, C., Bartsch, A., Tiemann, P., Reschechtko, S., Latash, M.L., 2018. Multi-finger synergies and the muscular apparatus of the hand. Experimental Brain Research 236, 1383–1393.

Cuadra, C., Latash, M.L., 2019. Exploring the concept of iso-perceptual manifold (IPM): a study of finger force matching. Neuroscience 401, 130–141.

Cuadra, C., Falaki, A., Sarlegna, F., Sainburg, R.L., Latash, M.L., 2019. Finger interaction and coordination in a deafferented person: implications for the neural control of the hand. In: Progress in Motor Control, July 7-11, 2019, Amsterdam: The Netherlands.

Curioni, A., Minio-Paluello, I., Sacheli, L.M., Candidi, M., Aglioti, S.M., 2017. Autistic traits affect interpersonal motor coordination by modulating strategic use of role-based behavior. Molecular Autism 8, 23.

Danion, F., Latash, M.L., Li, S., 2003a. Finger interactions studied with transcranial magnetic stimulation during multi-finger force production tasks. Clinical Neurophysiology 114, 1445–1455.

Danion, F., Schöner, G., Latash, M.L., Li, S., Scholz, J.P., Zatsiorsky, V.M., 2003b. A force mode hypothesis for finger interaction during multi-finger force production tasks. Biological Cybernetics 88, 91–98.

Danna-Dos-Santos, A., Degani, A.M., Zatsiorsky, V.M., Latash, M.L., 2008. Is voluntary control of natural postural sway possible? Journal of Motor Behavior 40, 179–185.

d'Avella, A., Saltiel, P., Bizzi, E., 2003. Combinations of muscle synergies in the construction of a natural motor behavior. Nature Neuroscience 6, 300–308.

Decety, J., 1996. Do imagined and executed actions share the same neural substrate? Brain Research: Cognitive Brain Research 3, 87–93.

De Freitas, P.B., Freitas, S.M.S.F., Lewis, M.M., Huang, X., Latash, M.L., 2018. Individual preferences in motor coordination seen across the two hands: relations to movement stability and optimality. Experimental Brain Research 237, 1–13.

De Gail, P., Lance, J.W., Neilson, P.D., 1966. Differential effects on tonic and phasic reflex mechanisms produced by vibration of muscles in man. Journal of Neurology Neurosurgery and Psychiatry 29, 1–11.

Degani, A.M., Leonard, C.T., Danna-Dos-Santos, A., 2017. The effects of early stages of aging on postural sway: a multiple domain balance assessment using a force platform. Journal of Biomechanics 64, 8–15.

De Havas, J., Ghosh, A., Gomi, H., Haggard, P., 2016. Voluntary motor commands reveal awareness and control of involuntary movement. Cognition 155, 155–167.

De Havas, J., Gomi, H., Haggard, P., 2017. Experimental investigations of control principles of involuntary movement: a comprehensive review of the Kohnstamm phenomenon. Experimental Brain Research 235, 1953–1997.

De Havas, J., Ito, S., Haggard, P., Gomi, H., 2018. Low gain servo control during the Kohnstamm phenomenon reveals dissociation between low-level control mechanisms for involuntary vs. voluntary arm movements. Frontiers in Behavioral Neuroscience 12, 113.

Delwaide, P.J., Toulouse, P., 1983. The Jendrassik maneuver: quantitative analysis of reflex reinforcement by remote voluntary muscle contraction. Advances in Neurology 39, 661–669.

Demeny, G., 1890. Du role mécanique des muscles antagonists. Archives de Physiologie Ser 5 (2), 747.

DeWald, J.P., Pope, P.S., Given, J.D., Buchanan, T.S., Rymer, W.Z., 1995. Abnormal muscle coactivation patterns during isometric torque generation at the elbow and shoulder in hemiparetic subjects. Brain 118, 495–510.

Diedrichsen, J., Shadmehr, R., Ivry, R.B., 2010. The coordination of movement: optimal feedback control and beyond. Trends in Cognitive Sciences 14, 31–39.

Dietz, V., 2000. Spastic movement disorder. Spinal Cord 38, 389–393.

Dietz, V., Berger, W., 1983. Normal and impaired regulation of muscle stiffness in gait: a new hypothesis about muscle hypertonia. Experimental Neurology 79, 680–687.

Dietz, V., Quintern, J., Berger, W., 1981. Electrophysiological studies of gait in spasticity and rigidity. Evidence that altered mechanical properties of muscle contribute to hypertonia. Brain 104, 431–449.

Dietz, V., Sinkjaer, T., 2012. Spasticity. Handbook of Clinical Neurology 109, 197–211.

Dijkstra, T.M., Schöner, G., Gielen, C.C., 1994. Temporal stability of the action-perception cycle for postural control in a moving visual environment. Experimental Brain Research 97, 477–486.

Dinse, H.R., 2006. Cortical reorganization in the aging brain. Progress in Brain Research 157, 57–80.

DiZio, P., Lackner, J.R., 1995. Motor adaptation to Coriolis force perturbations of reaching movements: endpoint but not trajectory adaptation transfers to the nonexposed arm. Journal of Neurophysiology 74, 1787–1792.

Duclos, C., Roll, R., Kavounoudias, A., Roll, J.P., 2007. Cerebral correlates of the "Kohnstamm phenomenon": an fMRI study. Neuroimage 34, 774–783.

Duysens, J., Trippel, M., Horstmann, G.A., Dietz, V., 1990. Gating and reversal of reflexes in ankle muscles during human walking. Experimental Brain Research 82, 351–358.

Earhart, G.M., Jones, G.M., Horak, F.B., Block, E.W., Weber, K.D., Fletcher, W.A., 2002. Podokinetic after-rotation following unilateral and bilateral podokinetic stimulation. Journal of Neurophysiology 87, 1138–1141.

Eklund, G., Hagbarth, K.E., 1966. Normal variability of tonic vibration reflexes in man. Experimental Neurology 16, 80–92.

Eklund, G., Hagbarth, K.E., 1967. Vibratory induced motor effects in normal man and in patients with spastic paralysis. Electroencephalography and Clinical Neurophysiology 23, 393.

Eisen, A., Entezari-Taher, M., Stewart, H., 1996. Cortical projections to spinal motoneurons: changes with aging and amyotrophic lateral sclerosis. Neurology 46, 1396–1404.

Erim, Z., Beg, F.M., Burke, D.T., De Luca, C.J., 1999. Effects of aging on motor-unit control properties. Journal of Neurophysiology 82, 2081–2091.

Erlhagen, W., Schöner, G., 2002. Dynamic field theory of movement preparation. Psychological Review 109, 545–572.

Ernst, M.O., 2009. Perceptual learning: inverting the size-weight illusion. Current Biology 19, R23–R25.

Evarts, E.V., 1968. Relation of pyramidal tract activity to force exerted during voluntary movement. Journal of Neurophysiology 31, 14–27.

Fadiga, L., Craighero, L., Olivier, E., 2005. Human motor cortex excitability during the perception of others' action. Current Opinion in Neurobiology 15, 213–218.

Fahn, S., Jankovic, J., 2007. Principle and Practice of Movement Disorders. Churchill, Livingstone, Philadelphia, PA.

Falaki, A., Huang, X., Lewis, M.M., Latash, M.L., 2016. Impaired synergic control of posture in Parkinson's patients without postural instability. Gait and Posture 44, 209–215.

Falaki, A., Huang, X., Lewis, M.M., Latash, M.L., 2017a. Dopaminergic modulation of multi-muscle synergies in postural tasks performed by patients with Parkinson's disease. Journal of Electromyography and Kinesiology 33, 20–26.

Falaki, A., Huang, X., Lewis, M.M., Latash, M.L., 2017b. Motor equivalence and structure of variance: multi-muscle postural synergies in Parkinson's disease. Experimental Brain Research 235, 2243–2258.

Falaki, A., Jo, H.J., Lewis, M.M., O'Connell, B., De Jesus, S., McInerney, J., Huang, X., Latash, M.L., 2018. Systemic effects of deep brain stimulation on synergic control in Parkinson's disease. Clinical Neurophysiology 129, 1320–1332.

Feldman, A.G., 1966. Functional tuning of the nervous system with control of movement or maintenance of a steady posture. II. Controllable parameters of the muscle. Biophysics 11, 565–578.

Feldman, A.G., 1980. Superposition of motor programs. I. Rhythmic forearm movements in man. Neuroscience 5, 81–90.

Feldman, A.G., 1986. Once more on the equilibrium-point hypothesis (λ-model) for motor control. Journal of Motor Behavior 18, 17–54.

Feldman, A.G., 2009. New insights into action-perception coupling. Experimental Brain Research 194, 39–58.

Feldman, A.G., 2015. Referent Control of Action and Perception: Challenging Conventional Theories in Behavioral Science. Springer, NY.

Feldman, A.G., 2016. Active sensing without efference copy: referent control of perception. Journal of Neurophysiology 116, 960–976.

Feldman, A.G., 2019. Indirect, referent control of motor actions underlies directional tuning of neurons. Journal of Neurophysiology 121, 823–841.

Feldman, A.G., Latash, M.L., 1982a. Interaction of afferent and efferent signals underlying joint position sense: empirical and theoretical approaches. Journal of Motor Behavior 14, 174–193.

Feldman, A.G., Latash, M.L., 1982b. Afferent and efferent components of joint position sense: interpretation of kinaesthetic illusions. Biological Cybernetics 42, 205–214.

Feldman, A.G., Latash, M.L., 1982c. Inversions of vibration-induced senso-motor events caused by supraspinal influences in man. Neuroscience Letters 31, 147–151.

Feldman, A.G., Latash, M.L., 2005. Testing hypotheses and the advancement of science: recent attempts to falsify the equilibrium-point hypothesis. Experimental Brain Research 161, 91–103.

Feldman, A.G., Levin, M.F., 1995. Positional frames of reference in motor control: their origin and use. Behavioral and Brain Sciences 18, 723–806.

Feldman, A.G., Orlovsky, G.N., 1972. The influence of different descending systems on the tonic stretch reflex in the cat. Experimental Neurology 37, 481–494.

Feldman, A.G., Ilmane, N., Sangani, S., Raptis, H., Esmailzadeh, N., 2013. Action-perception coupling in kinesthesia: a new approach. Neuropsychologia 51, 2590–2599.

Fetz, E.E., Cheney, P.D., 1980. Postspike facilitation of forelimb muscle activity by primate corticomotoneuronal cells. Journal of Neurophysiology 44, 751–772.

Flanagan, J.R., Johansson, R.S., 2002. Hand movements. In: Ramshandran, V.S. (Ed.), Encyclopaedia of the Human Brain. Academic Press, San Diego, pp. 399–414.

Flash, T., Hogan, N., 1985. The coordination of arm movements: an experimentally confirmed mathematical model. Journal of Neuroscience 5, 1688–1703.

Folkins, J.W., Linville, R.N., 1983. The effects of varying lower-lip displacement on upper-lip movements: implications for the coordination of speech movements. Journal of Speech and Hearing Research 26, 209–217.

Forssberg, H., 1979. Stumbling corrective reaction: a phase dependent compensatory reaction during locomotion. Journal of Neurophysiology 42, 936–953.

Forssberg, H., Grillner, S., Rossignol, S., 1975. Phase dependent reflex reversal during walking in chronic spinal cat. Brain Research 85, 103–107.

Fortier, P.A., Kalaska, J.F., Smith, A.M., 1989. Cerebellar neuronal activity related to whole-arm reaching movements in the monkey. Journal of Neurophysiology 62, 198–211.

Freitas, S.M.S.F., de Freitas, P.B., Lewis, M.M., Huang, X., Latash, M.L., 2019. Quantitative analysis of multi-element synergies stabilizing performance: comparison of three methods with respect to their use in clinical studies. Experimental Brain Research 237, 453–465.

Friedman, J., SKM, V., Zatsiorsky, V.M., Latash, M.L., 2009. The sources of two components of variance: an example of multifinger cyclic force production tasks at different frequencies. Experimental Brain Research 196, 263–277.

Frysinger, R.C., Bourbonnais, D., Kalaska, J.F., Smith, A.M., 1984. Cerebellar cortical activity during antagonist cocontraction and reciprocal inhibition of forearm muscles. Journal of Neurophysiology 51, 32–49.

Fuentes, C.T., Gomi, H., Haggard, P., 2012. Temporal features of human tendon vibration illusions. European Journal of Neuroscience 36, 3709–3717.

Fukson, O.I., Berkinblit, M.B., Feldman, A.G., 1980. The spinal frog takes into account the scheme of its body during the wiping reflex. Science 209, 1261–1263.

Gatts, S., 2008. Neural mechanisms underlying balance control in Tai Chi. Medical Sport Science 52, 87–103.

Gelfand, I.M., 1991. Two archetypes in the psychology of man. Nonlinear Science Today 1, 11–16.

Gelfand, I.M., Latash, M.L., 1998. On the problem of adequate language in movement science. Motor Control 2, 306–313.

Gelfand, I.M., Tsetlin, M.L., 1962. On certain methods of control of complex systems. Advances in Mathematical Sciences 17, 103 (in Russian).

Gelfand, I.M., Tsetlin, M.L., 1966. On mathematical modeling of the mechanisms of the central nervous system. In: Gelfand, I.M., Gurfinkel, V.S., Fomin, S.V., Tsetlin, M.L. (Eds.), Models of the Structural-Functional Organization of Certain Biological Systems. Nauka, Moscow, pp. 9–26 (in Russian, a translation is available in 1971 edition by MIT Press: Cambridge MA).

Georgopoulos, A.P., 1986. On reaching. Annual Review of Neuroscience 9, 147–170.

Georgopoulos, A.P., Ashe, J., Smyrnis, N., Taira, M., 1992. The motor cortex and the coding of force. Science 256, 1692–1695.

Georgopoulos, A.P., Kalaska, J.F., Caminiti, R., Massey, J.T., 1982. On the relations between the direction of two-dimensional arm movements and cell discharge in primate motor cortex. Journal of Neuroscience 2, 1527–1537.

Georgopoulos, A.P., Kalaska, J.F., Caminiti, R., Massey, J.T., 1983. Interruption of motor cortical discharge subserving aimed arm movements. Experimental Brain Research 49, 327–340.

Georgopoulos, A.P., Lurito, J.T., Petrides, M., Schwartz, A.B., Massey, J.T., 1989. Mental rotation of the neuronal population vector. Science 243, 234–236.

Georgopoulos, A.P., Schwartz, A.B., Kettner, R.E., 1986. Neural population coding of movement direction. Science 233, 1416–1419.

Gescheider, G.A., 1997. Psychophysics: The Fundamentals, third ed. Lawrence Erlbaum, New York, NY.

Ghez, C., Gordon, J., 1987. Trajectory control in targeted force impulses. I. Role of opposing muscles. Experimental Brain Research 67, 225–240.

Gibson, J.J., 1979. The Ecological Approach to Visual Perception. Houghton Mifflin, Boston, MA.

Glansdorf, P., Prigogine, I., 1971. Thermodynamic Theory of Structures, Stability and Fluctuations. Wiley.

Godfrey, S.B., Ajoudani, A., Catalano, M., Grioli, G., Bicchi, A., 2013. A synergy-driven approach to a myoelectric hand. In: IEEE International Conference on Rehabilitation Robotics, vol. 2013, p. 6650377.

Gomi, H., Kawato, M., 1996. Equilibrium-point hypothesis examined by measured arm stiffness during multijoint movement. Science 272, 117–120.
Goodwin, G.M., McCloskey, D.I., Matthews, P.B., 1972. The contribution of muscle afferents to kinaesthesia shown by vibration induced illusions of movement and by the effects of paralysing joint afferents. Brain 95, 705–748.
Gordon, C.R., Fletcher, W.A., Melvill Jones, G., Block, E.W., 1995. Adaptive plasticity in the control of locomotor trajectory. Experimental Brain Research 102, 540–545.
Gorniak, S., Zatsiorsky, V.M., Latash, M.L., 2007. Hierarchies of synergies: an example of the two-hand, multi-finger tasks. Experimental Brain Research 179, 167–180.
Gorniak, S., Zatsiorsky, V.M., Latash, M.L., 2009. Hierarchical control of static prehension: II. Multi-digit synergies. Experimental Brain Research 194, 1–15.
Gorniak, S.L., Zatsiorsky, V.M., Latash, M.L., 2011. Manipulation of a fragile object by elderly individuals. Experimental Brain Research 212, 505–516.
Gottlieb, G.L., 1998. Rejecting the equilibrium-point hypothesis. Motor Control 2, 10–12.
Gottlieb, G.L., Agarwal, G.C., 1986. The invariant characteristic isn't. Behavioral and Brain Sciences 9, 608–609.
Gottlieb, G.L., Agarwal, G.C., 1988. Compliance of single joints: elastic and plastic characteristics. Journal of Neurophysiology 59, 937–951.
Gottlieb, G.L., Corcos, D.M., Agarwal, G.C., 1989a. Strategies for the control of voluntary movements with one mechanical degree of freedom. Behavioral and Brain Sciences 12, 189–250.
Gottlieb, G.L., Corcos, D.M., Agarwal, G.C., 1989b. Organizing principles for single joint movements. I: a speed-insensitive strategy. Journal of Neurophysiology 62, 342–357.
Grabiner, M.D., Enoka, R.M., 1995. Changes in movement capabilities with aging. Exercise and Sport Sciences Reviews 23, 65–104.
Granit, R., 1970. The Basis of Motor Control. Academic Press, London.
Gregory, J.E., Wood, S.A., Proske, U., 2001. An investigation into mechanisms of reflex reinforcement by the Jendrassik manoeuvre. Experimental Brain Research 138, 366–374.
Gribble, P.L., Ostry, D.J., Sanguineti, V., Laboissiere, R., 1998. Are complex control signals required for human arm movements? Journal of Neurophysiology 79, 1409–1424.
Grigg, P., Finerman, G.A., Riley, L.H., 1973. Joint-position sense after total hip replacement. Journal of Bone and Joint Surgery, American 55, 1016–1025.
Gromov, M., 2018. Great Circles of Mysteries. Mathematics, the World, the Mind. Birkhäuser, Basel, Switzerland.
Gurfinkel, V.S., Levik, Y.S., Kazennikov, O.V., Selionov, V.A., 1998. Locomotor-like movements evoked by leg muscle vibration in humans. European Journal of Neuroscience 10, 1608–1612.
Hammond, M.C., Fitts, S.S., Kraft, G.H., Nutter, P.B., Trotter, M.J., Robinson, L.M., 1988. Co-contraction in the hemiparetic forearm: quantitative EMG evaluation. Archives of Physical Medicine and Rehabilitation 69, 348–351.
Harris, C.M., Wolpert, D.M., 1998. Signal-dependent noise determines motor planning. Nature 394, 780–784.
Harris, J.M., Best, C.S., Moffat, V.J., Spencer, M.D., Philip, R.C., Power, M.J., Johnstone, E.C., 2008. Autistic traits and cognitive performance in young people with mild intellectual impairment. Journal of Autism and Developmental Disorders 38, 1241–1249.
Hasan, Z., 1986. Optimized movement trajectories and joint stiffness in unperturbed, inertially loaded movements. Biological Cybernetics 53, 373–382.

Hasanbarani, F., Reschechtko, S., Latash, M.L., 2018. Performance drifts in two-finger cyclical force production tasks performed by one and two actors. Experimental Brain Research 236, 779–794.

Hayashi, R., Miyake, A., Jijiwa, H., Watanabe, S., 1981. Postural readjustment to body sway induced by vibration in man. Experimental Brain Research 43, 217–225.

Heckman, C.J., Gorassini, M.A., Bennett, D.J., 2005. Persistent inward currents in motoneuron dendrites: implications for motor output. Muscle and Nerve 31, 135–156.

Heckman, C.J., Johnson, M., Mottram, C., Schuster, J., 2008. Persistent inward currents in spinal motoneurons and their influence on human motoneuron firing patterns. The Neuroscientist 14, 264–275.

Heijnen, M.J., Muir, B.C., Rietdyk, S., 2012. Factors leading to obstacle contact during adaptive locomotion. Experimental Brain Research 223, 219–231.

Heijnen, M.J., Romine, N.L., Stumpf, D.M., Rietdyk, S., 2014. Memory-guided obstacle crossing: more failures were observed for the trail limb versus lead limb. Experimental Brain Research 232, 2131–2142.

Hepp-Reymond, M., Kirkpatrick-Tanner, M., Gabernet, L., Qi, H.X., Weber, B., 1999. Context-dependent force coding in motor and premotor cortical areas. Experimental Brain Research 128, 123–133.

Hinder, M.R., Milner, T.E., 2003. The case for an internal dynamics model versus equilibrium point control in human movement. Journal of Physiology 549, 953–963.

Hirai, H., Miyazaki, F., Naritomi, H., Koba, K., Oku, T., Uno, K., Uemura, M., Nishi, T., Kageyama, M., Krebs, H.I., 2015. On the origin of muscle mynergies: Invariant balance in the co-activation of agonist and antagonist muscle pairs. Frontiers in Bioengineering and Biotechnology 3, 192.

Hirokawa, S., Solomonow, M., Luo, Z., Lu, Y., D'Ambrosia, R., 1991. Muscular co-contraction and control of knee stability. Journal of Electromyography and Kinesiology 1, 199–208.

Hoehn, M., Yahr, M., 1967. Parkinsonism: onset, progression and mortality. Neurology 17, 427–442.

Hogan, N., 1984. An organizational principle for a class of voluntary movements. Journal of Neuroscience 4, 2745–2754.

Holdefer, R.N., Miller, L.E., 2002. Primary motor cortical neurons encode functional muscle synergies. Experimental Brain Research 146, 233–243.

Hömberg, V., Hefter, H., Reiners, K., Freund, H.J., 1987. Differential effects of changes in mechanical limb properties on physiological and pathological tremor. Journal of Neurology Neurosurgery and Psychiatry 50, 568–579.

Homma, S., Kanda, K., Watanabe, S., 1971. Tonic vibration reflex in human and monkey subjects. The Japanese Journal of Physiology 21, 419–430.

Hommel, B., Müsseler, J., Aschersleben, G., Prinz, W., 2001. The Theory of Event Coding (TEC): a framework for perception and action planning. Behavioral and Brain Sciences 24, 849–878 discussion 878-937.

Horak, F.B., Nutt, J.G., Nashner, L.M., 1992. Postural inflexibility in parkinsonian subjects. Journal of Neurological Sciences 111, 46–58.

Houk, J.C., 2005. Agents of the mind. Biological Cybernetics 92, 427–437.

Hsu, W.L., Scholz, J.P., Schöner, G., Jeka, J.J., Kiemel, T., 2007. Control and estimation of posture during quiet stance depends on multijoint coordination. Journal of Neurophysiology 97, 3024–3035.

Hughlings Jackson, J., 1889. On the comparative study of disease of the nervous system. British Medical Journal 355–362 (Aug. 17), 1889.

Hulliger, M., Nordh, E., Vallbo, A.B., 1982. The absence of position response in spindle afferent units from human finger muscles during accurate position holding. Journal of Physiology 322, 167–179.

Humphrey, D.R., Corrie, W.S., Rietz, R., 1978. Properties of the pyramidal tract neuron system within the precentral wrist and hand area of primate motor cortex. Journal of Physiology (Paris) 74, 215–226.

Humphrey, D.R., 1982. Separate cell systems in the motor cortex of the monkey for the control of joint movement and of joint stiffness. Electroencephalography and Clinical Neurophysiology – Supplement 36, 393–408.

Humphrey, D.R., Reed, D.J., 1983. Separate cortical systems for control of joint movement and joint stiffness: reciprocal activation and coactivation of antagonist muscles. Advances in Neurology 39, 347–372.

Humphrey, D.R., Schmidt, E.M., Thompson, W.D., 1970. Predicting measures of motor performance from multiple cortical spike trains. Science 170, 758–762.

Iggo, A., Muir, A.R., 1969. The structure and function of a slowly adapting touch corpuscle in hairy skin. Journal of Physiology 200, 763–796.

Ilmane, N., Sangani, S., Feldman, A.G., 2013. Corticospinal control strategies underlying voluntary and involuntary wrist movements. Behavioral and Brain Research 236, 350–358.

Imamizu, H., Uno, Y., Kawato, M., 1998. Adaptive internal model of intrinsic kinematics involved in learning an aiming task. Journal of Experimental Psychology: Human Perception and Performance 24, 812–829.

Ito, T., Kimura, T., Gomi, H., 2005. The motor cortex is involved in reflexive compensatory adjustment of speech articulation. NeuroReport 16, 1791–1794.

Ivanenko, Y.P., Poppele, R.E., Lacquaniti, F., 2004. Five basic muscle activation patterns account for muscle activity during human locomotion. Journal of Physiology 556, 267–282.

Ivanenko, Y.P., Wright, W.G., Gurfinkel, V.S., Horak, F., Cordo, P., 2006. Interaction of involuntary post-contraction activity with locomotor movements. Experimental Brain Research 169, 255–260.

Jaric, S., Collins, J.J., Marwaha, R., Russell, E., 2006. Interlimb and within limb force coordination in static bimanual manipulation task. Experimental Brain Research 168, 88–97.

Jacono, M., Casadio, M., Morasso, P.G., Sanguineti, V., 2004. The sway density curve and the underlying postural stabilization process. Motor Control 8, 292–311.

Jankowska, E., 1979. New observations on neuronal organization of reflexes from tendon organ afferents and their relation to reflexes evoked from muscle spindle afferents. In: Granit, R., Pompeiano, O. (Eds.), Reflex Control of Posture and Movement. Elsevier, Amsterdam, pp. 29–36.

Jankowska, E., McCrea, D.A., 1983. Shared reflex pathways from Ib tendon organ afferents and Ia muscle spindle afferents in the cat. Journal of Physiology 338, 99–111.

Jeannerod, M., Frak, V., 1999. Mental imaging of motor activity in humans. Current Opinion in Neurobiology 9, 735–739.

Jeka, J.J., Oie, K., Schoner, G., Dijkstra, T., Henson, E., 1998. Position and velocity coupling of postural sway to somatosensory drive. Journal of Neurophysiology 79, 1661–1674.

Jerde, T.E., Soechting, J.F., Flanders, M., 2003. Coarticulation in fluent fingerspelling. Journal of Neuroscience 23, 2383–2393.

Jo, H.J., Ambike, S., Lewis, M.M., Huang, X., Latash, M.L., 2016a. Finger force changes in the absence of visual feedback in patients with Parkinson's disease. Clinical Neurophysiology 127, 684–692.

Jo, H.J., Park, J., Lewis, M.M., Huang, X., Latash, M.L., 2015. Prehension synergies and hand function in early-stage Parkinson's disease. Experimental Brain Research 233, 425–440.

Jo, H.J., Maenza, C., Good, D.C., Huang, X., Park, J., Sainburg, R.L., Latash, M.L., 2016b. Effects of unilateral stroke on multi-finger synergies and their feed-forward adjustments. Neuroscience 319, 194–205.

Jo, H.J., Lucassen, E., Huang, X., Latash, M.L., 2017. Changes in multi-digit synergies and their feed-forward adjustments in multiple sclerosis. Journal of Motor Behavior 49, 218–228.

Jobin, A., Levin, M.F., 2000. Regulation of stretch reflex threshold in elbow flexors in children with cerebral palsy: a new measure of spasticity. Developmental Medicine and Child Neurology 42, 531–540.

Johansson, R.S., Westling, G., 1984. Roles of glabrous skin receptors and sensorimotor memory in automatic control of precision grip when lifting rougher or more slippery objects. Experimental Brain Research 56, 550–564.

Johnson, K., Ladefoged, P., Lindau, M., 1993. Individual differences in vowel production. Journal of the Acoustical Society of America 94, 701–714.

Kakei, S., Hoffman, D.S., Strick, P.L., 1999. Muscle and movement representations in the primary motor cortex. Science 285, 2136–2139.

Kakei, S., Hoffman, D.S., Strick, P.L., 2003. Sensorimotor transformations in cortical motor areas. Neuroscience Research 46, 1–10.

Kalaska, J.F., Cohen, D.A., Hyde, M.L., Prud'homme, M., 1989. A comparison of movement direction-related versus load direction-related activity in primate motor cortex, using a two- dimensional reaching task. Journal of Neuroscience 9, 2080–2102.

Kapur, S., Zatsiorsky, V.M., Latash, M.L., 2010. Age-related changes in the control of finger force vectors. Journal of Applied Physiology 109, 1827–1841.

Karanjia, P.N., Ferguson, J.H., 1983. Passive joint position sense after total hip replacement surgery. Annals of Neurology 13, 654–657.

Kawato, M., 1999. Internal models for motor control and trajectory planning. Current Opinion in Neurobiology 9, 718–727.

Kelso, J.A.S., 1995. Dynamic Patterns: The Self-Organization of Brain and Behavior. MIT Press, Cambridge.

Kelso, J.A., Holt, K.G., 1980. Exploring a vibratory systems analysis of human movement production. Journal of Neurophysiology 43, 1183–1196.

Kelso, J.A.S., Tuller, B., Fowler, C.A., 1982. On the functional specificity of articulatory control and coordination. Journal of the Acoustical Society of America 72, S103.

Kelso, J.A.S., Tuller, B., Vatikiotis-Bateson, E., Fowler, C.A., 1984. Functionally specific articulatory cooperation following jaw perturbations during speech: evidence for coordinative structures. Journal of Experimental Psychology: Human Perception and Performance 10, 812–832.

Kiemel, T., Oie, K.S., Jeka, J.J., 2002. Multisensory fusion and the stochastic structure of postural sway. Biological Cybernetics 87, 262–277.

Kilbreath, S.L., Gandevia, S.C., 1994. Limited independent flexion of the thumb and fingers in human subjects. Journal of Physiology 479, 487–497.

Kistemaker, D.A., Van Soest, A.J., Wong, J.D., Kurtzer, I., Gribble, P.L., 2013. Control of position and movement is simplified by combined muscle spindle and Golgi tendon organ feedback. Journal of Neurophysiology 109, 1126–1139.

Kitatani, R., Ohata, K., Sakuma, K., Aga, Y., Yamakami, N., Hashiguchi, Y., Yamada, S., 2016. Ankle muscle coactivation during gait is decreased immediately after anterior weight shift practice in adults after stroke. Gait and Posture 45, 35–40.

Klishko, A.N., Farrell, B.J., Beloozerova, I.N., Latash, M.L., Prilutsky, B.I., 2014. Stabilization of cat paw trajectory during locomotion. Journal of Neurophysiology 112, 1376–1391.

Körding, K.P., Wolpert, D.M., 2006. Bayesian decision theory in sensorimotor control. Trends in Cognitive Sciences 10, 319–326.

Krishnamoorthy, V., Goodman, S.R., Latash, M.L., Zatsiorsky, V.M., 2003a. Muscle synergies during shifts of the center of pressure by standing persons: identification of muscle modes. Biological Cybernetics 89, 152–161.

Krishnamoorthy, V., Latash, M.L., Scholz, J.P., Zatsiorsky, V.M., 2003b. Muscle synergies during shifts of the center of pressure by standing persons. Experimental Brain Research 152, 281–292.

Krishnamoorthy, V., Latash, M.L., Scholz, J.P., Zatsiorsky, V.M., 2004. Muscle modes during shifts of the center of pressure by standing persons: effects of instability and additional support. Experimental Brain Research 157, 18–31.

Krumhansl, C.L., 2000. Rhythm and pitch in music cognition. Psychological Bulletin 126, 159–179.

Kudo, K., Tsutsui, S., Ishikura, T., Ito, T., Yamamoto, Y., 2000. Compensatory coordination of release parameters in a throwing task. Journal of Motor Behavior 32, 337–345.

Kugler, P.N., Turvey, M.T., 1987. Information, Natural Law, and the Self-Assembly of Rhythmic Movement. Erlbaum, Hillsdale, NJ.

Lackner, J.R., DiZio, P., 1994. Rapid adaptation to Coriolis force perturbations of arm trajectory. Journal of Neurophysiology 72, 1–15.

Lackner, J.R., Levine, M.S., 1979. Changes in apparent body orientation and sensory localization, induced by vibration of postural muscles; vibratory myesthetic illusions. Aviation, Space, and Environmental Medicine 50, 346–354.

Lackner, J.R., Taublieb, A.B., 1984. Influence of vision on vibration-induced illusions of limb movement. Experimental Neurology 85, 97–106.

Lacquaniti, F., Licata, F., Soechting, J.F., 1982. The mechanical behavior of the human forearm in response to transient perturbations. Biological Cybernetics 44, 67–77.

Lafe, C.W., Pacheco, M.M., Newell, K.M., 2016a. Adapting relative phase of bimanual isometric force coordination through scaling visual information intermittency. Human Movement Science 47, 186–196.

Lafe, C.W., Pacheco, M.M., Newell, K.M., 2016b. Bimanual coordination and the intermittency of visual information in isometric force tracking. Experimental Brain Research 234, 2025–2034.

Lance, J.W., 1980. The control of muscle tone, reflexes, and movement: robert Wartenberg Lecture. Neurology 30, 1303–1313.

Landau, W.M., 1974. Spasticity: the fable of a neurological demon and the emperor's new therapy. Archives of Neurology 31, 217–219.

Lang, C.E., Schieber, M.H., 2003. Differential impairment of individuated finger movements in humans after damage to the motor cortex or the corticospinal tract. Journal of Neurophysiology 90, 1160–1170.

Latash, M.L., 1992. Virtual trajectories, joint stiffness, and changes in natural frequency during single-joint oscillatory movements. Neuroscience 49, 209–220.

Latash, M.L., 1993. Control of Human Movement. Human Kinetics, Urbana, IL.

Latash, M.L., 1994. Reconstruction of equilibrium trajectories and joint stiffness patterns during single-joint voluntary movements under different instructions. Biological Cybernetics 71, 441–450.

Latash, M.L., 1996. How does our brain make its choices? In: Latash, M.L., Turvey, M.T. (Eds.), Dexterity and its Development. Erlbaum Publ., Mahwah, NJ, pp. 277–304.

Latash, M.L., 2008. Synergy. Oxford University Press, New York.

Latash, M.L., 2010. Motor synergies and the equilibrium-point hypothesis. Motor Control 14, 294–322.

Latash, M.L., 2012a. The bliss (not the problem) of motor abundance (not redundancy). Experimental Brain Research 217, 1–5.

Latash, M.L., 2012b. Fundamentals of Motor Control. Academic Press, New York, NY.

Latash, M.L., 2016. Towards physics of neural processes and behavior. Neuroscience and Biobehavioral Reviews 69, 136–146.

Latash, M.L., 2017. Biological movement and laws of physics. Motor Control 21, 327–344.

Latash, M.L., 2018a. Stability of kinesthetic perception in efferent-afferent spaces: the concept of iso-perceptual manifold. Neuroscience 372, 97–113.

Latash, M.L., 2018b. Muscle co-activation: definitions, mechanisms, and functions. Journal of Neurophysiology 120, 88–104.

Latash, M.L., Anson, J.G., 1996. What are normal movements in atypical populations? Behavioral and Brain Sciences 19, 55–106.

Latash, M.L., Aruin, A.S., Zatsiorsky, V.M., 1999. The basis of a simple synergy: reconstruction of joint equilibrium trajectories during unrestrained arm movements. Human Movement Science 18, 3–30.

Latash, M.L., Friedman, J., Kim, S.W., Feldman, A.G., Zatsiorsky, V.M., 2010. Prehension synergies and control with referent hand configurations. Experimental Brain Research 202, 213–229.

Latash, M.L., Gottlieb, G.L., 1990. Compliant characteristics of single joints: preservation of equifinality with phasic reactions. Biological Cybernetics 62, 331–336.

Latash, M.L., Gottlieb, G.L., 1991. Reconstruction of elbow joint compliant characteristics during fast and slow voluntary movements. Neuroscience 43, 697–712.

Latash, M.L., Gottlieb, G.L., 1992. Virtual trajectories of single-joint movements performed under two basic strategies. Neuroscience 47, 357–365.

Latash, M.L., Gurfinkel, V.S., 1976. Tonic vibration reflex and position of the body. Physiologiya Cheloveka (Human Physiology) 2, 593–598.

Latash, M.L., Gutman, S.R., 1994. Abnormal motor patterns in the framework of the equilibrium-point hypothesis: a cause for dystonic movements? Biological Cybernetics 71, 87–94.

Latash, M.L., Kang, N., Patterson, D., 2002a. Finger coordination in persons with down syndrome: atypical patterns of coordination and the effects of practice. Experimental Brain Research 146, 345–355.

Latash, M.L., Penn, R.D., 1996. Changes in voluntary motor control induced by intrathecal baclofen. Physiotherapy Research International 1, 229–246.

Latash, M.L., Penn, R.D., Corcos, D.M., Gottlieb, G.L., 1989. Short-term effects of intrathecal baclofen in spasticity. Experimental Neurology 103, 165–172.

Latash, M.L., Penn, R.D., Corcos, D.M., Gottlieb, G.L., 1990. Effects of intrathecal baclofen on voluntary motor control in spastic paresis. Journal of Neurosurgery 72, 388–392.

Latash, M.L., Scholz, J.F., Danion, F., Schöner, G., 2001. Structure of motor variability in marginally redundant multi-finger force production tasks. Experimental Brain Research 141, 153–165.

Latash, M.L., Scholz, J.F., Danion, F., Schöner, G., 2002b. Finger coordination during discrete and oscillatory force production tasks. Experimental Brain Research 146, 412–432.

Latash, M.L., Scholz, J.P., Schöner, G., 2007. Toward a new theory of motor synergies. Motor Control 11, 276–308.

Latash, M.L., Shim, J.K., Smilga, A.V., Zatsiorsky, V., 2005. A central back-coupling hypothesis on the organization of motor synergies: a physical metaphor and a neural model. Biological Cybernetics 92, 186–191.

Latash, M.L., Huang, X., 2015. Neural control of movement stability: lessons from studies of neurological patients. Neuroscience 301, 39–48.

Latash, M.L., Zatsiorsky, V.M., 1993. Joint stiffness: myth or reality? Human Movement Science 12, 653–692.

Latash, M.L., Zatsiorsky, V.M., 2016. Biomechanics and Motor Control: Defining Central Concepts. Academic Press, New York, NY.

Lebedev, M.A., Nicolelis, M.A., 2017. Brain-machine interfaces: from basic science to neuroprostheses and neurorehabilitation. Physiological Reviews 97, 767–837.

Lee, D.N., Young, D.S., Rewt, D., 1992. How do somersaulters land on their feet? Journal of Experimental Psychology: Human Perception and Performance 18, 1195–1202.

Lee, P.J., Rogers, E.L., Granata, K.P., 2006. Active trunk stiffness increases with co-contraction. Journal of Electromyography and Kinesiology 16, 51–57.

Lee, Y., Ashton-Miller, J.A., 2011. The effects of gender, level of co-contraction, and initial angle on elbow extensor muscle stiffness and damping under a step increase in elbow flexion moment. Annals of Biomedical Engineering 39, 2542–2549.

Lee, Y.J., Chen, B., Aruin, A.S., 2015. Older adults utilize less efficient postural control when performing pushing task. Journal of Electromyography and Kinesiology 25, 966–972.

Lemon, R.N., Mantel, G.W., 1989. The influence of changes in discharge frequency of corticospinal neurones on hand muscles in the monkey. Journal of Physiology 413, 351–378.

Leone, F.C., Nottingham, R.B., Nelson, L.S., 1961. The folded normal distribution. Technometrics 3, 543–550.

Lestienne, F., 1979. Effects of inertial load and velocity on the braking process of voluntary limb movements. Experimental Brain Research 35, 407–418.

Levin, M.F., Feldman, A.G., 1994. The role of stretch reflex threshold regulation in normal and impaired motor control. Brain Research 657, 23–30.

Levin, O., Wenderoth, N., Steyvers, M., Swinnen, S.P., 2003. Directional invariance during loading-related modulations of muscle activity: evidence for motor equivalence. Experimental Brain Research 148, 62–76.

Levy-Tzedek, S., Ben Tov, M., Karniel, A., 2011. Rhythmic movements are larger and faster but with the same frequency on removal of visual feedback. Journal of Neurophysiology 106, 2120–2126.

Lewis, M.M., Lee, E.-Y., Jo, H.J., Park, J., Latash, M.L., Huang, X., 2016. Synergy as a new and sensitive marker of basal ganglia dysfunction: a study of asymptomatic welders. Neurotoxicology 56, 76–85.

Liddell, E.G.T., Sherrington, C.S., 1924. Reflexes in response to stretch (myotatic reflexes). Proceedings of the Royal Society of London, Series B 96, 212–242.

Lim, C.L., Yiannikas, C., 1992. Motor evoked potentials: a new method of controlled facilitation using quantitative surface EMG. Electroencephalography and Clinical Neurophysiology 85, 38–41.

Lingam, R., Golding, J., Jongmans, M.J., Hunt, L.P., Ellis, M., Emond, A., 2010. The association between developmental coordination disorder and other developmental traits. Pediatrics 126 (5), e1109–e1118.

Lisin, V.V., Frankstein, S.I., Rechtman, M.B., 1973. The influence of locomotion on flexor reflex of the hind limb in cat and man. Experimental Neurology 38, 180–183.

Liu, H., Hu, Z., Peng, D., 2013. Evaluation word in phrase: the modulation effect of emotional context on word comprehension. Journal of Psycholinguistic Research 42, 379–391.

Llinás, R., Welsh, J.P., 1993. On the cerebellum and motor learning. Current Opinion in Neurobiology 3, 958–965.
Loeb, G.E., 2012. Optimal isn't good enough. Biological Cybernetics 106, 757–765.
Lundberg, A., 1979. Multisensory control of spinal reflex pathways. In: Granit, R., Pompeiano, O. (Eds.), Reflex Control of Posture and Movement. Elsevier, Amsterdam, pp. 11–28.
Luu, B.L., Day, B.L., Cole, J.D., Fitzpatrick, R.C., 2011. The fusimotor and reafferent origin of the sense of force and weight. Journal of Physiology 589, 3135–3147.
MacKay, W.A., Crammond, D.J., Kwan, H.C., Murphy, J.T., 1986. Measurements of human forearm viscoelasticity. Journal of Biomechanics 19, 231–238.
MacKenzie, C.L., Iberall, T., 1994. The Grasping Hand. North-Holland, Amsterdam; New York.
Magnus, R., 1924. Korperstellung. Springer, Berlin.
Maier, M.A., Bennett, K.M., Hepp-Reymond, M.C., Lemon, R.N., 1983. Contribution of the monkey corticomotoneuronal system to the control of force in precision grip. Journal of Neurophysiology 69, 772–785.
Mangalam, M., Rein, R., Fragaszy, D.M., 2018. Bearded capuchin monkeys use joint synergies to stabilize the hammer trajectory while cracking nuts in bipedal stance. Proceedings. of the Royal Society. Biological Sciences 285 (1889).
Marsden, C.D., Rothwell, J.C., Traub, M., 1979. Long latency stretch reflex of the human thumb can be reversed if the task is changed. Journal of Physiology 293, 41P–42P.
Martin, J.R., Budgeon, M.K., Zatsiorsky, V.M., Latash, M.L., 2011. Stabilization of the total force in multi-finger pressing tasks studied with the "inverse piano" technique. Human Movement Science 30, 446–458.
Martin, V., Scholz, J.P., Schöner, G., 2009. Redundancy, self-motion, and motor control. Neural Computation 21, 1371–1414.
Massaro, D.W., 2004. From multisensory integration to talking heads and language learning. In: Calvert, G., Spence, C., Stein, B.E. (Eds.), The Handbook of Multisensory Processes. MIT Press, Cambridge, MA, pp. 153–176.
Massion, J., 1992. Movement, posture and equilibrium – interaction and coordination. Progress in Neurobiology 38, 35–56.
Matthews, P.B.C., 1959. The dependence of tension upon extension in the stretch reflex of the soleus of the decerebrate cat. Journal of Physiology 47, 521–546.
Matthews, P.B., Stein, R.B., 1969. The sensitivity of muscle spindle afferents to small sinusoidal changes of length. Journal of Physiology 200, 723–743.
Mathis, J., Gurfinkel, V.S., Struppler, A., 1996. Facilitation of motor evoked potentials by postcontraction response (Kohnstamm phenomenon). Electroencephalography and Clinical Neurophysiology 101, 289–297.
Mattos, D., Kuhl, J., Scholz, J.P., Latash, M.L., 2013. Motor equivalence (ME) during reaching: is ME observable at the muscle level? Motor Control 17, 145–175.
Mattos, D., Latash, M.L., Park, E., Kuhl, J., Scholz, J.P., 2011. Unpredictable elbow joint perturbation during reaching results in multijoint motor equivalence. Journal of Neurophysiology 106, 1424–1436.
Mattos, D., Schöner, G., Zatsiorsky, V.M., Latash, M.L., 2015. Motor equivalence during accurate multi-finger force production. Experimental Brain Research 233, 487–502.
Maurer, C., Peterka, R.J., 2005. A new interpretation of spontaneous sway measures based on a simple model of human postural control. Journal of Neurophysiology 93, 189–200.
McGurk, H., MacDonald, J., 1976. Hearing lips and seeing voices. Nature 264, 746–748.
McLaughlin, P., 2001. What Functions Explain. Cambridge University Press, Cambridge, UK.

McShea, D.W., Brandon, R.N., 2010. Biology's First Law. The University of Chicago Press, Chicago, London.

Mechsner, F., Kerzel, D., Knoblich, G., Prinz, W., 2001. Perceptual basis of bimanual coordination. Nature 414 (6859), 69–73.

Mel'cuk, I.A., 1981. Meaning-text models: a recent trend in Soviet linguistics. Annual Review of Anthropology 10, 27–62.

Merton, P.A., 1953. Speculations on the servo-control of movements. In: Malcolm, J.L., Gray, J.A.B., Wolstenholm, G.E.W. (Eds.), The Spinal Cord. Little, Brown, Boston, pp. 183–198.

Michalska, J., Kamieniarz, A., Fredyk, A., Bacik, B., Juras, G., Słomka, K.J., 2018. Effect of expertise in ballet dance on static and functional balance. Gait and Posture 64, 68–74.

Miller, L.E., Saygin, A.P., 2013. Individual differences in the perception of biological motion: links to social cognition and motor imagery. Cognition 128, 140–148.

Milner, T.E., Cloutier, C., 1993. Compensation for mechanically unstable loading in voluntary wrist movement. Experimental Brain Research 94, 522–532.

Misono, Y., Mazuka, R., Kondo, T., Kiritani, S., 1997. Effects and limitations of prosodic and semantic biases on syntactic disambiguation. Journal of Psycholinguistic Research 26, 229–245.

Mizuguchi, N., Kanosue, K., 2017. Changes in brain activity during action observation and motor imagery: their relationship with motor learning. Progress in Brain Research 234, 189–204.

Mochizuki, L., Duarte, M., Amadio, A.C., Zatsiorsky, V.M., Latash, M.L., 2006. Changes in postural sway and its fractions in conditions of postural instability. Journal of Applied Biomechanics 22, 51–66.

Mohan, V., Bhat, A., Morasso, P., 2018. Muscleless motor synergies and actions without movements: from motor neuroscience to cognitive robotics. Physics of Life Reviews. https://doi.org/10.1016/j.plrev.2018.04.005 (e-pub ahead of print).

Monteiro Ferronato, P.A., Barela, J.A., 2011. Age-related changes in postural control: rambling and trembling trajectories. Motor Control 15, 481–493.

Müller, H., Sternad, D., 2003. A randomization method for the calculation of covariation in multiple nonlinear relations: illustrated with the example of goal-directed movements. Biological Cybernetics 89, 22–33.

Müller, H., Sternad, D., 2004. Decomposition of variability in the execution of goal-oriented tasks: Three components of skill improvement. Journal of Experimental Psychology: Human Perception and Performance 30, 212–233.

Mullick, A.A., Musampa, N.K., Feldman, A.G., Levin, M.F., 2013. Stretch reflex spatial threshold measure discriminates between spasticity and rigidity. Clinical Neurophysiology 124, 740–751.

Mullick, A.A., Turpin, N.A., Hsu, S.C., Subramanian, S.K., Feldman, A.G., Levin, M.F., 2018. Referent control of the orientation of posture and movement in the gravitational field. Experimental Brain Research 236, 381–398.

Munhall, K.G., Gribble, P., Sacco, L., Ward, M., 1996. Temporal constraints on the McGurk effect. Perception and Psychophysics 58, 351–362.

Mussa Ivaldi, F.A., 1988. Do neurons in the motor cortex encode movement direction? An alternative hypothesis. Neuroscience Letters 91, 106–111.

Nakajima, T., Kamibayashi, K., Kitamura, T., Komiyama, T., Zehr, E.P., Nakazawa, K., 2016. Short-term plasticity in a monosynaptic reflex pathway to forearm muscles after continuous robot-assisted passive stepping. Frontiers in Human Neuroscience 10, 368.

Nasir, S.M., Ostry, D.J., 2006. Somatosensory precision in speech production. Current Biology 16, 1918–1923.
Nelson, W., 1983. Physical principles for economies of skilled movements. Biological Cybernetics 46, 135–147.
Nichols, T.R., 1994. A biomechanical perspective on spinal mechanisms of coordinated muscular action: an architecture principle. Acta Anatomica 151, 1–13.
Nichols, T.R., 2002. Musculoskeletal mechanics: a foundation of motor physiology. Advances in Experimental Medicine and Biology 508, 473–479.
Nichols, T.R., Houk, J.C., 1976. Improvement in linearity and regulation of stiffness that results from actions of stretch reflex. Journal of Neurophysiology 39, 119–142.
Nichols, T.R., Steeves, J.D., 1986. Resetting of resultant stiffness in ankle flexor and extensor muscles in the decerebrate cat. Experimental Brain Research 62, 401–410.
Nielsen, J.B., Kagamihara, Y., 1992. The regulation of disynaptic reciprocal Ia inhibition during co-contraction of antagonistic muscles in man. Journal of Physiology 456, 373–391.
Nielsen, J.B., Crone, C., Hultborn, H., 2007. The spinal pathophysiology of spasticity—from a basic science point of view. Acta Physiologica (Oxford) 189, 171–180.
Nieto-Castanon, A., Guenther, F.H., Perkell, J.S., Curtin, H.D., 2005. A modeling investigation of articulatory variability and acoustic stability during American english/r/production. Journal of the Acoustical Society of America 117, 3196–3212.
Niu, X., Latash, M.L., Zatsiorsky, V.M., 2012. Reproducibility and variability of the cost functions reconstructed from experimental recordings in multi-finger prehension. Journal of Motor Behavior 44, 69–85.
Olafsdottir, H., Kim, S.W., Zatsiorsky, V.M., Latash, M.L., 2008. Anticipator synergy adjustments in preparation to self-triggered perturbations in elderly individuals. Journal of Applied Biomechanics 24, 175–179.
Olafsdottir, H., Yoshida, N., Zatsiorsky, V.M., Latash, M.L., 2005. Anticipatory covariation of finger forces during self-paced and reaction time force production. Neuroscience Letters 381, 92–96.
Olafsdottir, H., Yoshida, N., Zatsiorsky, V.M., Latash, M.L., 2007a. Elderly show decreased adjustments of motor synergies in preparation to action. Clinical Biomechanics 22, 44–51.
Olatsdottir, H., Zatsiorsky, V.M., Latash, M.L., 2005. Is the thumb a fifth finger? A study of digit interaction during force production tasks. Experimental Brain Research 160, 203–213.
Olafsdottir, H., Zhang, W., Zatsiorsky, V.M., Latash, M.L., 2007b. Age related changes in multi-finger synergies in accurate moment of force production tasks. Journal of Applied Physiology 102, 1490–1501.
Ostry, D.J., Feldman, A.G., 2003. A critical evaluation of the force control hypothesis in motor control. Experimental Brain Research 153, 275–288.
Pandarinath, C., Ames, K.C., Russo, A.A., Farshchian, A., Miller, L.E., Dyer, E.L., Kao, J.C., 2018. Latent factors and dynamics in motor cortex and their application to brain-machine interfaces. Journal of Neuroscience 38, 9390–9401.
Park, J., Jo, H.J., Lewis, M.M., Huang, X., Latash, M.L., 2013b. Effects of Parkinson's disease on optimization and structure of variance in multi-finger tasks. Experimental Brain Research 231, 51–63.
Park, J., Lewis, M.M., Huang, X., Latash, M.L., 2013a. Effects of olivo-ponto-cerebellar atrophy (OPCA) on finger interaction and coordination. Clinical Neurophysiology 124, 991–998.
Park, J., Lewis, M.M., Huang, X., Latash, M.L., 2014. Dopaminergic modulation of motor coordination in Parkinson's disease. Parkinsonism and Related Disorders 20, 64–68.
Park, J., Singh, T., Zatsiorsky, V.M., Latash, M.L., 2012b. Optimality vs. variability: effect of fatigue in multi-finger redundant tasks. Experimental Brain Research 216, 591–607.

Park, J., Wu, Y.-H., Lewis, M.M., Huang, X., Latash, M.L., 2012a. Changes in multi-finger interaction and coordination in Parkinson's disease. Journal of Neurophysiology 108, 915–924.

Park, J., Sun, Y., Zatsiorsky, V.M., Latash, M.L., 2011. Age-related changes in optimality and motor variability: an example of multi-finger redundant tasks. Experimental Brain Research 212, 1–18.

Park, J., Zatsiorsky, V.M., Latash, M.L., 2010. Optimality vs. variability: an example of multi-finger redundant tasks. Experimental Brain Research 207, 119–132.

Parsa, B., O'Shea, D.J., Zatsiorsky, V.M., Latash, M.L., 2016. On the nature of unintentional action: a study of force/moment drifts during multi-finger tasks. Journal of Neurophysiology 116, 698–708.

Parsa, B., Terekhov, A.V., Zatsiorsky, V.M., Latash, M.L., 2017. Optimality and stability of intentional and unintentional actions: I. Origins of drifts in performance. Experimental Brain Research 235, 481–496.

Partridge, L.D., 1965. Modifications of neural output signals by muscles: a frequency response study. Journal of Applied Physiology 20, 150–156.

Pawlowitsch, C., Mertikopoulos, P., Ritt, N., 2011. Neutral stability, drift, and the diversification of languages. Journal of Theoretical Biology 287, 1–12.

Penfield, W., Rasmussen, T., 1950. The Cerebral Cortex of Man. A Clinical Study of Localization of Function. MacMillan, New York.

Penn, R.D., Savoy, S., Corcos, D., Latash, M., Gottlieb, G., Parke, B., Kroin, J., 1989. Intrathecal baclofen for severe spinal spasticity: A double-blind crossover study. New England Journal of Medicine 320, 1517–1521.

Perkell, J.S., Matthies, M.L., Svirsky, M.A., Jordan, M.I., 1993. Trading relations between tongue-body raising and lip rounding in production of the vowel/u/: a pilot "motor equivalence" study. Journal of the Acoustical Society of America 93, 2948–2961.

Peterka, R.J., 2002. Sensorimotor integration in human postural control. Journal of Neurophysiology 88, 1097–1118.

Petrosini, L., Cutuli, D., Picerni, E., Laricchiuta, D., 2015. Cerebellum and personality traits. The Cerebellum 14, 43–46.

Phillips, C.G., 1969. Motor apparatus of the baboon's hand. Proceedings of the Royal Society of London, Series B 173, 141–174.

Picerni, E., Petrosini, L., Piras, F., Laricchiuta, D., Cutuli, D., Chiapponi, C., Fagioli, S., Girardi, P., Caltagirone, C., Spalletta, G., 2013. New evidence for the cerebellar involvement in personality traits. Frontiers in Behavioral Neuroscience 7, 133.

Poliakov, A.V., Schieber, M.H., 1999. Limited functional grouping of neurons in the motor cortex hand area during individuated finger movements: a cluster analysis. Journal of Neurophysiology 82, 3488–3505.

Polit, A., Bizzi, E., 1978. Processes controlling arm movements in monkey. Science 201, 1235–1237.

Polit, A., Bizzi, E., 1979. Characteristics of motor programs underlying arm movement in monkey. Journal of Neurophysiology 42, 183–194.

Pompeiano, O., Andre, P., Manzoni, D., 1997. Spatiotemporal response properties of cerebellar Purkinje cells to animal displacement: a population analysis. Neuroscience 81, 609–626.

Poon, C., Chin-Cottongim, L.G., Coombes, S.A., Corcos, D.M., Vaillancourt, D.E., 2012. Spatiotemporal dynamics of brain activity during the transition from visually guided to memory-guided force control. Journal of Neurophysiology 108, 1335–1348.

Popescu, F.C., Rymer, W.Z., 2000. End points of planar reaching movements are disrupted by small force pulses: an evaluation of the hypothesis of equifinality. Journal of Neurophysiology 84, 2670–2679.
Prilutsky, B.I., Zatsiorsky, V.M., 2002. Optimization-based models of muscle coordination. Exercise and Sport Sciences Reviews 30, 32–38.
Prochazka, A., Clarac, F., Loeb, G.E., Rothwell, J.C., Wolpaw, J.R., 2000. What do reflex and voluntary mean? Modern views on an ancient debate. Experimental Brain Research 130, 417–432.
Prochazka, A., Gritsenko, V., Yakovenko, S., 2002. Sensory control of locomotion: reflexes versus higher-level control. Advances in Experimental Medicine and Biology 508, 357–367.
Proske, U., Gandevia, S.C., 2012. The proprioceptive senses: their roles in signaling body shape, body position and movement, and muscle force. Physiological Reviews 92, 1651–1697.
Proverbio, A.M., Massetti, G., Rizzi, E., Zani, A., 2016. Skilled musicians are not subject to the McGurk effect. Scientific Reports 6, 30423.
Rabinovich, M.I., Simmons, A.N., Varona, P., 2015. Dynamical bridge between brain and mind. Trends in Cognitive Sciences 19, 453–461.
Raczaszek, J., Tuller, B., Shapiro, L.P., Case, P., Kelso, S., 1999. Categorization of ambiguous sentences as a function of a changing prosodic parameter: a dynamical approach. Journal of Psycholinguistic Research 28, 367–393.
Ralston, H.J., Lukin, L., 1969. Energy levels of human body segments during level walking. Ergonomics 12, 39–46.
Rand, M.K., Wang, L., Müsseler, J., Heuer, H., 2013. Vision and proprioception in action monitoring by young and older adults. Neurobiology of Aging 34, 1864–1872.
Raptis, H., Burtet, L., Forget, R., Feldman, A.G., 2010. Control of wrist position and muscle relaxation by shifting spatial frames of reference for motoneuronal recruitment: possible involvement of corticospinal pathways. Journal of Physiology 588, 1551–1570.
Raptis, H.A., Dannenbaum, E., Paquet, N., Feldman, A.G., 2007. Vestibular system may provide equivalent motor actions regardless of the number of body segments involved in the task. Journal of Neurophysiology 97, 4069–4078.
Rasouli, O., Solnik, S., Furmanek, M.P., Piscitelli, D., Falaki, A., Latash, M.L., 2017. Unintentional drifts during quiet stance and voluntary body sway. Experimental Brain Research 235, 2301–2316.
Ravnborg, M., Blinkenberg, M., Dahl, K., 1991. Standardization of facilitation of compound muscle action potentials evoked by magnetic stimulation of the cortex. Electroencephalography and Clinical Neurophysiology 81, 195–201.
Reinkensmeyer, D.J., Akoner, O., Ferris, D.P., Gordon, K.E., 2009. Slaking by the human motor system: computational models and implications for robotic orthosis. IEEE Engineering in Medicine and Biology Society 2009, 2129–2132.
Reimann, H., Schöner, G., 2017. A multi-joint model of quiet, upright stance accounts for the "uncontrolled manifold" structure of joint variance. Biological Cybernetics 111, 389–403.
Reisman, D., Scholz, J.P., 2003. Aspects of joint coordination are preserved during pointing in persons with post-stroke hemiparesis. Brain 126, 2510–2527.
Reschechtko, S., Cuadra, C., Latash, M.L., 2018. Force illusions and drifts observed during muscle vibration. Journal of Neurophysiology 119, 326–336.
Reschechtko, S., Hasanbarani, F., Akulin, V.M., Latash, M.L., 2017a. Unintentional force changes in cyclical tasks performed by an abundant system: empirical observations and a dynamical model. Neuroscience 350, 94–109.

Reschechtko, S., Latash, M.L., 2017. Stability of hand force production: I. Hand level control variables and multi-finger synergies. Journal of Neurophysiology 118, 3152–3164.

Reschechtko, S., Latash, M.L., 2018. Stability of hand force production: II. Ascending and descending synergies. Journal of Neurophysiology 120, 1045–1060.

Reschechtko, S., Zatsiorsky, V.M., Latash, M.L., 2014. Stability of multi-finger action in different spaces. Journal of Neurophysiology 112, 3209–3218.

Reschechtko, S., Zatsiorsky, V.M., Latash, M.L., 2015. Task-specific stability of multi-finger steady-state action. Journal of Motor Behavior 47, 365–377.

Reschechtko, S., Zatsiorsky, V.M., Latash, M.L., 2017b. The synergic control of multi-finger force production: stability of explicit and implicit task components. Experimental Brain Research 235, 1–14.

Resnick, B., 2015. Study: Elite Scientists Can Hold Back Science. Vox. https://www.vox.com/science-and-health/2015/12/15/10219330/elite-scientists-hold-back-progress.

Riccio, G.E., 1993. Information in movement variability about the qualitative dynamics of posture and orientation. In: Newell, K.M., Corcos, D.M. (Eds.), Variability and Motor Control. Human Kinetics Publ., Champaign, IL, pp. 317–358.

Richards, C.L., Malouin, F., 2013. Cerebral palsy: definition, assessment and rehabilitation. Handbook of Clinical Neurology 111, 183–195.

Richardson, M.J., Shockley, K., Fajen, B.R., Riley, M.A., Turvey, M.T., 2008. Ecological psychology: six principles for an embodied-embedded approach to behavior. In: Calvo, P., Gomila, T. (Eds.), Handbook of Cognitive Science: An Embodied Approach. Elsevier, New York, NY, pp. 161–187.

Rizzolatti, G., Craighero, L., 2004. The mirror-neuron system. Annual Review of Neuroscience 27, 169–192.

Robilliard, J.J., Pfau, T., Wilson, A.M., 2007. Gait characterisation and classification in horses. Journal of Experimental Biology 210, 187–197.

Roll, J.P., Vedel, J.P., 1982. Kinaesthetic role of muscle afferents in man, studied by tendon vibration and microneurography. Experimental Brain Research 47, 177–190.

Rosenbaum, D.A., Engelbrecht, S.E., Busje, M.M., Loukopoulos, L.D., 1993. Knowledge model for selecting and producing reaching movements. Journal of Motor Behavior 25, 217–227.

Rossini, P.M., Barker, A.T., Berardelli, A., Caramia, M.T., Caruso, G., Gracco, R.Q., Dimitrijevic, M.R., Hallett, M., Katayama, Y., Lucking, C.H., Maertens-de Noordhout, A.L., Marsden, C.D., Murray, N.M.F., Rothwell, J.C., Swash, M., Tomberg, C., 1994. Non-invasive electrical and magnetic stimulation of the brain, spinal cord and roots; basic principles and procedures for routine clinical application. Electroencephalography and Clinical Neurophysiology 91, 79–92.

Russell, D.M., Haworth, J.L., Martinez-Garza, C., 2016. Coordination dynamics of (a)symmetrically loaded gait. Experimental Brain Research 234, 867–881.

Russo, A.A., Bittner, S.R., Perkins, S.M., Seely, J.S., London, B.M., Lara, A.H., Miri, A., Marshall, N.J., Kohn, A., Jessell, T.M., Abbott, L.F., Cunningham, J.P., Churchland, M.M., 2018. Motor cortex embeds muscle-like commands in an untangled population response. Neuron 97, 953–966.

Rymer, W.Z., D'Almeida, A., 1980. Joint position sense. The effects of muscle contraction. Brain 103, 1–22.

Sainburg, R.L., 2002. Evidence for a dynamic-dominance hypothesis of handedness. Experimental Brain Research 142, 241–258.

Sainburg, R.L., 2005. Handedness: differential specializations for control of trajectory and position. Exercise and Sport Sciences Reviews 33, 206–213.

Sainburg, R.L., Poizner, H., Ghez, C., 1995. Loss of proprioception produces deficits in interjoint coordination. Journal of Neurophysiology 70, 2136–2147.
Sainburg, R.L., Ghilardi, M.F., Poizner, H., Ghez, C., 1993. Control of limb dynamics in normal subjects and patients without proprioception. Journal of Neurophysiology 73, 820–835.
Sangani, S.G., Raptis, H.A., Feldman, A.G., 2011. Subthreshold corticospinal control of anticipatory actions in humans. Behavioral and Brain Research 224, 145–154.
Santello, M., Flanders, M., Soechting, J.F., 2002. Patterns of hand motion during grasping and the influence of sensory guidance. Journal of Neuroscience 22, 1426–1435.
Sarabon, N., Panjan, A., Latash, M.L., 2013. The effects of aging on the rambling and trembling components of postural sway: effects of motor and sensory challenges. Gait and Posture 38, 637–642.
Sarabon, N., Rosker, J., 2013. Effect of 14 days of bed rest in older adults on parameters of the body sway and on the local ankle function. Journal of Electromyography and Kinesiology 23, 1505–1511.
Schieber, M.H., Santello, M., 2004. Hand function: peripheral and central constraints on performance. Journal of Applied Physiology 96, 2293–2300.
Schmidt, R.A., 1975. A schema theory of discrete motor skill learning. Psychological Review 82, 225–260.
Schmidt, R.A., 1980. Past and future issues in motor programming. Research Quarterly for Exercise and Sport 51, 122–140.
Schmidt, R.A., McGown, C., 1980. Terminal accuracy of unexpected loaded rapid movements: evidence for a mass-spring mechanism in programming. Journal of Motor Behavior 12, 149–161.
Schmidt, R.C., Carello, C., Turvey, M.T., 1990. Phase transitions and critical fluctuations in the visual coordination of rhythmic movements between people. Journal of Experimental Psychology: Human Perception and Performance 16, 227–247.
Schmidt, R.C., Turvey, M.T., 1994. Phase-entrainment dynamics of visually coupled rhythmic movements. Biological Cybernetics 170, 369–376.
Scholz, J.P., Danion, F., Latash, M.L., Schöner, G., 2002. Understanding finger coordination through analysis of the structure of force variability. Biological Cybernetics 86, 29–39.
Scholz, J.P., Kang, N., Patterson, D., Latash, M.L., 2003. Uncontrolled manifold analysis of single trials during multi-finger force production by persons with and without Down syndrome. Experimental Brain Research 153, 45–58.
Scholz, J.P., Schöner, G., 1999. The uncontrolled manifold concept: identifying control variables for a functional task. Experimental Brain Research 126, 289–306.
Scholz, J.P., Schöner, G., Hsu, W.L., Jeka, J.J., Horak, F., Martin, V., 2007. Motor equivalent control of the center of mass in response to support surface perturbations. Experimental Brain Research 180, 163–179.
Scholz, J.P., Schöner, G., Latash, M.L., 2000. Identifying the control structure of multijoint coordination during pistol shooting. Experimental Brain Research 135, 382–404.
Schöner, G., 1995. Recent developments and problems in human movement science and their conceptual implications. Ecological Psychology 8, 291–314.
Schöner, G., Kelso, J.A.S., 1988. Dynamic pattern generation in behavioral and neural systems. Science 239, 1513–1520.
Schöner, G., Thelen, E., 2006. Using dynamic field theory to rethink infant habituation. Psychological Review 113, 273–299.
Schrödinger, E., 1944, 2012. What is Life? Canto Classics. Cambridge University Press, Cambridge, UK.

Scott, J.T., Lohnes, C.A., Horak, F.B., Earhart, G.M., 2011. Podokinetic stimulation causes shifts in perception of straight ahead. Experimental Brain Research 208, 313–321.

Scott, S.H., Kalaska, J.F., 1995. Changes in motor cortex activity during reaching movements with similar hand paths but different arm postures. Journal of Neurophysiology 73, 2563–2567.

Searle, J.R., 1992. The Rediscovery of the Mind. A Bradford Book. MIT Edition, Boston, MA.

Seif-Naraghi, A.H., Winters, J.M., 1990. Optimized strategies for scaling goal-directed dynamic limb movements. In: Winters, J.M., Woo, S.L.-Y. (Eds.), Multiple Muscle Systems. Biomechanics and Movement Organization. Springer-Verlag, New York e.a., pp. 312–334

Sergio, L.E., Kalaska, J.F., 1998. Changes in the temporal pattern of primary motor cortex activity in a directional isometric force versus limb movement task. Journal of Neurophysiology 80, 1577–1583.

Sergio, L.E., Hamel-Pâquet, C., Kalaska, J.F., 2005. Motor cortex neural correlates of output kinematics and kinetics during isometric-force and arm-reaching tasks. Journal of Neurophysiology 94, 2353–2378.

Shaiman, S., 2002. Articulatory control of vowel length for contiguous jaw cycles: the effects of speaking rate and phonetic context. Journal of Speech, Language, and Hearing Research 45, 663–675.

Shin, S., Motl, R.W., Sosnoff, J.J., 2011. A test of the rambling and trembling hypothesis: multiple sclerosis and postural control. Motor Control 15, 568–579.

Shin, S., Sosnoff, J.J., 2017. Spinal cord injury and seated postural control: a test of the rambling and trembling hypothesis. Motor Control 21, 443–456.

Seidler-Dobrin, R.D., Stelmach, G.E., 1998. Persistence in visual feedback control by the elderly. Experimental Brain Research 119, 467–474.

Selionov, V.A., Ivanenko, Y.P., Solopova, I.A., Gurfinkel, V.S., 2009. Tonic central and sensory stimuli facilitate involuntary air-stepping in humans. Journal of Neurophysiology 101, 2847–2858.

Sergio, L.E., Kalaska, J.F., 1997. Systematic changes in directional tuning of motor cortex cell activity with hand location in the workspace during generation of static isometric forces in constant spatial directions. Journal of Neurophysiology 78, 1170–1174.

Sittig, A.C., Denier van der Gon, J.J., Gielen, C.C., 1985. Separate control of arm position and velocity demonstrated by vibration of muscle tendon in man. Experimental Brain Research 60, 445–453.

Shadmehr, R., Wise, S.P., 2005. The Computational Neurobiology of Reaching and Pointing. MIT Press, Cambridge, MA.

Shim, J.K., Lay, B., Zatsiorsky, V.M., Latash, M.L., 2004a. Age-related changes in finger coordination in static prehension tasks. Journal of Applied Physiology 97, 213–224.

Shim, J.K., Olafsdottir, H., Zatsiorsky, V.M., Latash, M.L., 2005b. The emergence and disappearance of multi-digit synergies during force production tasks. Experimental Brain Research 164, 260–270.

Shinohara, M., Latash, M.L., Zatsiorsky, V.M., 2003a. Age effects on force produced by intrinsic and extrinsic hand muscles and finger interaction during MVC tasks. Journal of Applied Physiology 95, 1361–1369.

Shinohara, M., Li, S., Kang, N., Zatsiorsky, V.M., Latash, M.L., 2003b. Effects of age and gender on finger coordination in MVC and sub-maximal force-matching tasks. Journal of Applied Physiology 94, 259–270.

Shinohara, M., Scholz, J.P., Zatsiorsky, V.M., Latash, M.L., 2004. Finger interaction during accurate multi-finger force production tasks in young and elderly persons. Experimental Brain Research 156, 282–292.

Shiratori, T., Latash, M.L., 2000. The roles of proximal and distal muscles in anticipatory postural adjustments under asymmetrical perturbations and during standing on rollerskates. Clinical Neurophysiology 111, 613–623.

Slifkin, A.B., Vaillancourt, D.E., Newell, K.M., 2000. Intermittency in the control of continuous force production. Journal of Neurophysiology 84, 1708, 171.

Slobounov, S., Chiang, H., Johnston, J., Ray, W., 2002. Modulated cortical control of individual fingers in experienced musicians: an EEG study. Electroencephalographic study. Clinical Neurophysiology 113, 2013–2024.

Smith, A.M., 1981. The coactivation of antagonist muscles. Canadian Journal of Physiology and Pharmacology 59, 733–747.

Smith, B.W., Rowe, J.B., Reinkensmeyer, D.J., 2018. Real-time slacking as a default mode of grip force control: implications for force minimization and personal grip force variation. Journal of Neurophysiology 120, 2107–2120.

Solnik, S., Qiao, M., Latash, M.L., 2017. Effects of visual feedback and memory on unintentional drifts in performance during finger pressing tasks. Experimental Brain Research 235, 1149–1162.

Solopova, I.A., Selionov, V.A., Zhvansky, D.S., Gurfinkel, V.S., Ivanenko, Y., 2016. Human cervical spinal cord circuitry activated by tonic input can generate rhythmic arm movements. Journal of Neurophysiology 115, 1018–1030.

Sozzi, S., Schieppati, M., 2016. Stepping in place while voluntarily turning around produces a long-lasting posteffect consisting in inadvertent turning while stepping with eyes closed. Neural Plasticity 2016, 7123609.

Speedtsberg, M.B., Christensen, S.B., Andersen, K.K., Bencke, J., Jensen, B.R., Curtis, D.J., 2017. Impaired postural control in children with developmental coordination disorder is related to less efficient central as well as peripheral control. Gait and Posture 51, 1–6.

Sperry, R.W., 1950. Neural basis of the spontaneous optokinetic response produced by visual inversion. Journal of Comparative Physiology and Psychology 43, 482–489.

Stepp, N., Turvey, M.T., 2010. On strong anticipation. Cognitive Systems Research 11, 148–164.

Stepp, N., Turvey, M.T., 2017. Anticipation in manual tracking with multiple delays. Journal of Experimental Psychology: Human Perception and Performance 43, 914–925.

Sternad, D., 1998. A dynamic systems perspective to perception and action. Research Quarterly for Exercise and Sport 69, 319–325.

Sternad, D., 2001. Kurt Wachholder: pioneering electrophysiological studies of voluntary movements. In: Latash, M.L., Zatsiorsky, V.M. (Eds.), Classics in Movement Science. Human Kinetics, Urbana, IL, pp. 375–407.

Sternad, D., Collins, D., Turvey, M.T., 1995. The detuning factor in the dynamics of interlimb rhythmic coordination. Biological Cybernetics 73, 27–35.

Sternad, D., Duarte, M., Katsumata, H., Schaal, S., 2001. Bouncing a ball: tuning into dynamic stability. Journal of Experimental Psychology: Human Perception and Performance 27, 1163–1184.

Sternad, D., Turvey, M.T., Saltzman, E.L., 1999. Dynamics of 1:2 coordination: sources of symmetry breaking. Journal of Motor Behavior 31, 224–235.

Subramanian, S.K., Feldman, A.G., Levin, M.F., 2018. Spasticity may obscure motor learning ability after stroke. Journal of Neurophysiology 119, 5–20.

Suzuki, M., Yamazaki, Y., 2005. Velocity-based planning of rapid elbow movements expands the control scheme of the equilibrium point hypothesis. Journal of Computational Neuroscience 18, 131–149.

Tal'nov, A.N., Kostiukov, A.I., 1991. The manifestation of the hysteresis effects of muscle contraction in the cortically evoked coactivation of muscle antagonists. Neirofiziologiia 23, 481–484.

Terekhov, A.V., Pesin, Y.B., Niu, X., Latash, M.L., Zatsiorsky, V.M., 2010. An analytical approach to the problem of inverse optimization: an application to human prehension. Journal of Mathematical Biology 61, 423–453.

Tilney, F., Pike, F.H., 1925. Muscular Coordination Experimentally Studied in its Relation to the Cerebellum Archives of Neurology and Psychiatry, vol. 13, pp. 289–334.

Thompson, A.K., Wolpaw, J.R., 2015. Restoring walking after spinal cord injury: operant conditioning of spinal reflexes can help. The Neuroscientist 21, 203–215.

Todorov, E., 2004. Optimality principles in sensorimotor control. Nature Neuroscience 7, 907–915.

Todorov, E., Jordan, M.I., 2002. Optimal feedback control as a theory of motor coordination. Nature Neuroscience 5, 1226–1235.

Togo, S., Imamizu, H., 2016. Anticipatory synergy adjustments reflect individual performance of feedforward force control. Neuroscience Letters 632, 192–198.

Traverse, E., Lebon, F., Martin, A., 2018. Corticospinal and spinal excitabilities are modulated during motor imagery associated with somatosensory electrical nerve stimulation. Neural Plasticity 2018, 8265427.

Tremblay, S., Shiller, D.M., Ostry, D.J., 2003. Somatosensory basis of speech production. Nature 423, 866–869.

Tresch, M.C., Jarc, A., 2009. The case for and against muscle synergies. Current Opinion in Neurobiology 19, 601–607.

Turpin, N.A., Feldman, A.G., Levin, M.F., 2017. Stretch-reflex threshold modulation during active elbow movements in post-stroke survivors with spasticity. Clinical Neurophysiology 128, 1891–1897.

Turpin, N.A., Levin, M.F., Feldman, A.G., 2016. Implicit learning and generalization of stretch response modulation in humans. Journal of Neurophysiology 115, 3186–3194.

Turvey, M.T., 1990. Coordination. American Psychologist 45, 938–953.

Turvey, M.T., 2007. Action and perception at the level of synergies. Human Movement Science 26, 657–697.

Uno, Y., Kawato, M., Suzuki, R., 1989. Formation and control of optimal trajectory in human multijoint arm movement. Biological Cybernetics 61, 89–101.

Vaillancourt, D.E., Russell, D.M., 2002. Temporal capacity of short-term visuomotor memory in continuous force production. Experimental Brain Research 145, 275–285.

Vaillancourt, D.E., Thulborn, K.R., Corcos, D.M., 2003. Neural basis for the processes that underlie visually guided and internally guided force control in humans. Journal of Neurophysiology 90, 3330–3340.

Vaillancourt, D.E., Slifkin, A.B., Newell, K.M., 2001. Visual control of isometric force in Parkinson's disease. Neurophysiologia 39, 1410–1418.

Vallbo, A., 1970. Discharge patterns in human muscle spindle afferents during isometric voluntary contractions. Acta Physiologica Scandinavica 80, 552–566.

Van Doren, C.L., 1998. Differential effects of load stiffness on matching pinch force, finger span, and effort. Experimental Brain Research 120, 487–495.

Von Holst, E., Mittelstaedt, H., 1950/1973. The reafference principle. In: The Behavioral Physiology of Animals and Man. The Collected Papers of Erich von Holst, vol. 1. Martin R (translator) University of Miami Press, Coral Gables, Florida, pp. 139–173.

Von Uexkull, J., 1904. Die ersten Ursachen des Rhythmus in der Tierreihe. Ergebnisse der Physiologie 3, 1 (cited after Magnus 1924).

Wada, M., Kawahara, H., Shimada, S., Miyazaki, T., Baba, H., 2002. Joint proprioception before and after total knee arthroplasty. Clinical Orthopaedics and Related Research 403, 161–167.

Watson, J.D., Colebatch, J.G., McCloskey, D., 1984. Effects of externally imposed elastic loads on the ability to estimate position and force. Behavioral and Brain Research 13, 267–271.

Weaver, F.M., Follett, K., Stern, M., Hur, K., Harris, C., Marks, W.J., Rothlind, J., Sagher, O., Reda, D., Moy, C.S., Pahwa, R., Burchiel, K., Hogarth, P., Lai, E.C., Duda, J.E., Holloway, K., Samii, A., Horn, S., Bronstein, J., Stoner, G., Heemskerk, J., Huang, G.D., 2009. Bilateral deep brain stimulation vs. best medical therapy for patients with advanced Parkinson disease: a randomized controlled trial. Journal of American Medical Association 301, 63–73.

Weber, K.D., Fletcher, W.A., Gordon, C.R., Melvill Jones, G., Block, E.W., 1998. Motor learning in the 'podokinetic' system and its role in spatial orientation during locomotion. Experimental Brain Research 120, 377–385.

Weiler, J., Gribble, P.L., Pruszynski, J.A., 2019. Spinal stretch reflexes support efficient hand control. Nature Neuroscience 22, 529–533. https://doi.org/10.1038/s41593-019-0336-0 (in press).

Weiss, Y., Simoncelli, E.P., Adelson, E.H., 2002. Motion illusions as optimal percepts. Nature Neuroscience 5, 598–604.

Welford, A.T., 1984. Between bodily changes and performance: some possible reasons for slowing with age. Experimental Aging Research 10, 73–88.

Whitall, J., Clark, J.E., 2018. A perception-action approach to understanding typical and atypical motor development. Advances in Child Development and Behavior 55, 245–272.

Whitney, D.E., 1969. Resolved motion rate control of manipulators and human prostheses. IEEE Transactions on Man-Machine Systems 10, 47–53.

Wilhelm, L.A., Martin, J.A., Latash, M.L., Zatsiorsky, V.M., 2014. Finger enslaving in the dominant and non-dominant hand. Human Movement Science 33, 185–193.

Wilhelm, L., Zatsiorsky, V.M., Latash, M.L., 2013. Equifinality and its violations in a redundant system: Multi-finger accurate force production. Journal of Neurophysiology 110, 1965–1973.

Wilson, A.D., Bingham, G.P., Craig, J.C., 2003. Proprioceptive perception of phase variability. Journal of Experimental Psychology: Human Perception and Performance 29, 1179–1190.

Wilson, A.D., Snapp-Childs, W., Coats, R., Bingham, G.P., 2010. Learning a coordinated rhythmic movement with task-appropriate coordination feedback. Experimental Brain Research 205, 513–520.

Wilson, A.D., Bingham, G.P., 2008. Identifying the information for the visual perception of relative phase. Perception and Psychophysics 270, 465–476.

Wilson, T.D., Brekke, N., 1994. Mental contamination and mental correction: unwanted influences on judgments and evaluations. Psychological Bulletin 116, 117–142.

Wing, A.M., 2000. Motor control: mechanisms of motor equivalence in handwriting. Current Biology 10, R245–R248.

Winter, D.A., Prince, F., Frank, J.S., Powell, C., Zabjek, K.F., 1996. Unified theory regarding A/P and M/L balance in quiet stance. Journal of Neurophysiology 75, 2334–2343.

Wolpaw, J.R., 2007. Spinal cord plasticity in acquisition and maintenance of motor skills. Acta Physiologica (Oxford) 189, 155–169.

Wolpaw, J.R., 2018. The negotiated equilibrium model of spinal cord function. Journal of Physiology 596, 3469–3491.

Wolpaw, J.R., Bedlack, R.S., Reda, D.J., Ringer, R.J., Banks, P.G., Vaughan, T.M., Heckman, S.M., McCane, L.M., Carmack, C.S., Winden, S., McFarland, D.J., Sellers, E.W., Shi, H., Paine, T., Higgins, D.S., Lo, A.C., Patwa, H.S., Hill, K.J., Huang, G.D., Ruff, R.L., 2018. Independent home use of a brain-computer interface by people with amyotrophic lateral sclerosis. Neurology 91, e258–e267.

Wolpaw, J.R., Carp, J.S., 1993. Adaptive plasticity in spinal cord. Advances in Neurology 59, 163–174.

Wolpaw, J.R., Tennissen, A.M., 2001. Activity-dependent spinal cord plasticity in health and disease. Annual Review of Neuroscience 24, 807–843.

Wolpert, D.M., Miall, R.C., Kawato, M., 1998. Internal models in the cerebellum. Trends in Cognitive Sciences 2, 338–347.

Xin, X., Liu, Y., 2014. Control Design and Analysis for Underactuated Robotic Systems. Springer, NY.

Xu, Y., Terekhov, A.V., Latash, M.L., Zatsiorsky, V.M., 2012. Forces and moments generated by the human arm: variability and control. Experimental Brain Research 223, 159–175.

Yamagata, M., Falaki, A., Latash, M.L., 2019a. Effects of voluntary agonist-antagonist co-activation on stability of vertical posture. Motor Control 23, 304–326.

Yamagata, M., Popow, M., Latash, M.L., 2019b. Beyond rambling and trembling: effects of visual feedback on slow postural drift. Experimental Brain Research 237, 865–871.

Yang, J.-F., Scholz, J.P., 2005. Learning a throwing task is associated with differential changes in the use of motor abundance. Experimental Brain Research 163, 137–158.

Yousif, N., Cole, J., Rothwell, J., Diedrichsen, J., 2015. Proprioception in motor learning: lessons from a deafferented subject. Experimental Brain Research 233, 2449–2459.

Zaback, M., Cleworth, T.W., Carpenter, M.G., Adkin, A.L., 2015. Personality traits and individual differences predict threat-induced changes in postural control. Human Movement Science 40, 393–409.

Zatsiorsky, V.M., 1998. Kinematics of Human Motion. Human Kinetics, Champaign, IL.

Zatsiorsky, V.M., 2002. Kinetics of Human Motion. Human Kinetics, Champaign, IL.

Zatsiorsky, V.M., Duarte, M., 1999. Instant equilibrium point and its migration in standing tasks: rambling and trembling components of the stabilogram. Motor Control 3, 28–38.

Zatsiorsky, V.M., Duarte, M., 2000. Rambling and trembling in quiet standing. Motor Control 4, 185–200.

Zatsiorsky, V.M., Gao, F., Latash, M.L., 2003. Finger force vectors in multi-finger prehension. Journal of Biomechanics 36, 1745–1749.

Zatsiorsky, V.M., Gregory, R.W., Latash, M.L., 2002. Force and torque production in static multi-finger prehension: biomechanics and control. Part II. Control. Biological Cybernetics 87, 40–49.

Zatsiorsky, V.M., Latash, M.L., 2008. Multi-finger prehension: an overview. Journal of Motor Behavior 40, 446–476.

Zatsiorsky, V.M., Li, Z.M., Latash, M.L., 2000. Enslaving effects in multi-finger force production. Experimental Brain Research 131, 187–195.

Zhang, L., Feldman, A.G., Levin, M.F., 2018. Vestibular and corticospinal control of human body orientation in the gravitational field. Journal of Neurophysiology 120, 3026–3041.

Zhang, W., Sainburg, R.L., Zatsiorsky, V.M., Latash, M.L., 2006. Hand dominance and multi-finger synergies. Neuroscience Letters 409, 200–204.

Zhou, T., Solnik, S., Wu, Y.-H., Latash, M.L., 2014a. Equifinality and its violations in a redundant system: control with referent configurations in a multi-joint positional task. Motor Control 18, 405–424.

Zhou, T., Solnik, S., Wu, Y.-H., Latash, M.L., 2014b. Unintentional movements produced by back-coupling between the actual and referent body configurations: violations of equifinality in multi-joint positional tasks. Experimental Brain Research 232, 3847–3859.

Zhou, T., Zatsiorsky, V.M., Latash, M.L., 2015a. Unintentional changes in the apparent stiffness of the multi-joint limb. Experimental Brain Research 233, 2989–3004.

Zhou, T., Zhang, L., Latash, M.L., 2015b. Intentional and unintentional multi-joint movements: their nature and structure of variance. Neuroscience 289, 181–193.

Zhou, T., Zhang, L., Latash, M.L., 2015c. Characteristics of unintentional movements by a multi-joint effector. Journal of Motor Behavior 47, 352–361.

Index

Note: 'Page numbers followed by "f" indicate figures.'

D

E

FSC
www.fsc.org
MIX
Papier | Fördert
gute Waldnutzung
FSC® C083411

Kolibri 360 GmbH
Ferdinand-Jühlke-Straße 7
99095 Erfurt, Deutschland
produktsicherheit@kolibri360.de